PROTECTING STRUCTURES IN THE WILDLAND/URBAN INTERFACE

BY
DAVID H. COWARDIN

STUDENT HANDOUTS

TABLE OF CONTENTS

WILDLAND/URBAN INTERFACE FIRE TERMINOLOGY

TOPIC: WILDLAND/URBAN INTERFACE FIRE TACTICS AND STRATEGY
GLOSSARY OF TERMS

INTRODUCTION:

Wildland firefighting has its own jargon like other aspects of the fire service. The terms presented in this information sheet are by no mean complete. Yet, they provide the firefighter with a basic set of need-to-know terms to improve communications and safety when working in the wildland/urban interface fire environment.

AREA IGNITION:	Usually occurs in narrow canyons. Radiant heat preheats fuels causing release of combustible gases throughout the exposed area. When ignited this preheated fuel practically explodes.
ANCHOR POINT:	A point to start a fire attack. Usually at the heel, on a flank, or at a barrier. Also called the toe.
AVAILABLE FUEL:	Those fuel which will burn during passage of a flaming front under specific burning conditions.
*BACKFIRING:	Intentionally setting fire to fuels inside the control line to contain a rapidly spreading fire when an attack is indirect. Burning out an area to protect structures.
BARRIER:	Any obstruction to the spread of fire, typically an area or strip devoid of flammable fuel.
BLACK:	(See Burn)

*BLOW-UP:	Sudden increase in fire intensity or rate of spread sufficient to preclude direct control or to upset existing control plans. Often accompanied by violent convection and may have other characteristics of a firestorm.
*BURN:	The areas that the fire has consumed or blackened.
*BURNING OUT:	Intentionally setting fire to fuels inside the control line to strengthen the line when attack is direct, or parallel, with the control line tied at points of the fire. The control line is incomplete unless there is not any fuel between the fire and the line.
CANOPY:	The stratum containing crown of the tallest vegetation (living or dead) usually above 20 feet in height.
*CHIMNEY:	A Drainage running straight down from a ridge.
*CHUTE:	Drainage running down from a ridge at something less than a ninety degree angle.
COLD TRAILING:	A method of controlling a partly dead fire edge by carefully inspecting and feeling with the hand to detect any fire, digging out every live spot, and trenching any live edge.
CONFUSION AREA:	An area along ridges where opposing winds often meet causing directional changes in fire spread.
CONFLAGRATION:	A raging, destructive fire often used to denote a fire with a moving front as distinguished from a firestorm.

CONVERGENCE:	Net horizontal in-flow of air into a layer if vertical motion results at the surface. Associated with low-pressure system.
CONVECTION:	In meteorology, atmospheric motions that are predominantly vertical; i.e., usually means upward as opposed to subsidence (downward).
CONVECTION COLUMN:	The thermally produced ascending column of gases, smoke, and debris produced by a fire. Note: On multiple headed fires, more than one convection column may be present.
*COUNTER FIRE:	Fire set between main fire and a backfire to hasten the spread of the backfire. Also called a draft fire. The act of setting counter fires is sometimes called front firing or strip firing.
*CREEPING:	Fire burning with a low flame and spreading slowly.
*CRITICAL INCIDENT DAYS:	The number of day a year an agency could experience a major wildland/urban fire.
*CROWN FIRE:	A fire that advances from top-to-top of trees or shrubs more or less independently of the surface fire.
*CROWNING OUT:	Fire burning principally as a surface fire that intermittently ignites the crowns of trees or shrubs as it advances. Also called torching.
CUMULONIMBUS:	The ultimate growth of a cumulus cloud into a mushroom shape with considerable vertical growth, usually fibrous ice crystals tops, and probably accompanied by lightning, thunder, hail, and strong winds.
CUMULUS:	A principal low cloud type in the form of individual cells of sharp non-fibrous outline, and vertical development.

DEFENSIBLE SPACE:	A cleared area around a house of 30 to 100 feet.
DIRECT ATTACK:	A method of suppression that treats the fire as a whole, or all its burning edges, by wetting, cooling, smothering, or chemically quenching the fire or by mechanically separating the fire from unburned fuel.
DOWN HILL ATTACK RULES:	Eight rules for downhill fire attack or control activities.
ESCAPE ROUTE:	The egress path to be followed by firefighters when moving to a safety zone.
ESCARPMENT:	A steep slope separating two comparatively level sloping surfaces.
EXTREME FIRE BEHAVIOR:	A level of wildfire behavior characteristics that ordinarily precludes methods of direct control action. One or more of the following is usually involved: High rates-of-spread; prolific crowning and/or spotting; presence of convection column. Predictability is difficult because such fires often exercise some degree of influence on their environment; behaving erratically, sometimes dangerously.
FINE FUELS:	Fuels such as grass, leaves, draped pine needles, fern, tree moss, and some types of slash which ignite readily and are consumed rapidly when dry. Also called flash fuels.
FINGER:	An area extending away from a fire line creating an elongated burn pattern.
*FIRE BRAND:	Any source of heat natural or manmade that is capable of igniting wildland fuels. Flaming or glowing fuel particles can be carried naturally by wind, convection currents, or by gravity into unburned fuels.

FIRE BREAK:	A natural or constructed barrier utilized to stop or check fires that may occur or to provide a control line from which to work. Sometimes called a fire lane.
FIRE DANGER RATING:	The integration of fuel, site, weather and risk factors that affect the inception and behavior of wildfires.
FIRE EDGE:	The area between the burn and the green or unburned material.
FIRE LINE:	The part of a control line that is scraped or dug to mineral soil. Sometimes called a fire trail
FIRE MODEL:	A computer generated picture of a wildfire spread based on fuel type, wind, and topography.
FIRING OUT:	(SEE BURNING OUT)
*FIRESTORM:	Violent convection caused by a large continuous area of intense fire. Often characterized by destructively violent surface in drafts near and beyond the perimeter, and sometimes by tornado-like whirls.
FIRE WHIRL:	A spinning, moving column of ascending air rising from a vortex and carrying aloft smoke, debris and flames. These range from a foot or two in diameter to small tornadoes in size and intensity.
FLAME LENGTH:	The length of flames measured along their axis at the fire front. Flame length is an indicator of fire intensity.
FLAMING FRONT:	That zone of a moving fire within which the combustion is primarily burning. Light fuels typically have a shallow front; whereas heavy fuels have a deeper front.

FLANKING:	Attacking a fire by working along the sides (flanks) either simultaneously or successively from a less active or anchor point and endeavoring to connect the two lines at the head.
*FLANKS OF A FIRE:	The parts of a fire perimeter that are roughly parallel to the main direction of spread.
*FLARE-UP:	Any sudden acceleration of fire spread or intensification of the fire. Unlike blow-up, a flare-up is of relatively short duration and does not radically change existing control plans.
FLASHOVER:	Rapid combustion and/or explosion of unburned gases trapped at some distance from the main fire front. Usually occurs in poorly ventilated topography. More commonly associated with structural fire behavior. Could be called a wildland backdraft or smoke explosion.
FOEHN:	A dry wind with strong downward component, characteristic of mountainous regions. It is usually, but not always, warm for the season. May be called by various names such as Santa Ana, Mono, Chinook, Northerner, etc.
FREE BURNING:	The condition of a fire or part of a fire that has not been checked by natural barriers or by control measures.
FRONT:	A transition zone between two air masses of different densities.
FUEL:	The oven dry weight of all existing fuels in a given area. Loading or mass per unit area is usually expressed in tons per acre.

FUEL MODELING:	A simulated fuel complex in which all of the fuel is described appropriately for a mathematical fire spread model.
FUEL TYPE:	An identifiable association of fuel elements of distinctive species, form, size, arrangement, or other characteristics that will cause a predictable rate of fire spread or difficulty of control under specified weather conditions.
*GENERAL WIND:	Free air or large-scale wind caused by high and low pressure systems.
*GRADIENT WIND:	A wind that flows parallel to the isobars or contours. Does not occur at low elevations, but is realized at a height of roughly 1,500 feet above mean terrain height.
*GRAVITY WIND:	A wind directed down a slope caused by greater air density near the slope than at the same height at a distance from the slope. Also called a drainage or down-slope wind.
GREEN:	The unburned material in a wildfire path.
GROUND FIRE:	Fire that consumes the organic material beneath the surface litter, such as a peat fire.
HEAD OF A FIRE:	The most rapidly spreading portion of a fire perimeter, usually to the leeward or up-slope.
*HEADER:	Same as a convection column. Often used to describe a convection column seen from a distance.
HEEL OF A FIRE:	As distinguished from the head, the starting point on leeward or up-slope spreading fires. Usually having little, if any, active burning.

INDIRECT ATTACK: A method of suppression in which the control line is located along natural firebreaks, favorable breaks in topography, or at considerable distance from the fire. The intervening fuel is backfired or burned out.

ISLAND: An area inside the burn area that is unburned.

*INVERSION: An increase in temperature with height; i.e., a departure from the usual decrease in temperature with increase in altitude.

*LADDER FUELS: Fuels which provide vertical continuity between strata. Fire is able to carry from surface fuels through convection into the crown with relative ease.

LCES: Lookouts - Communications - Escape Route - Safety Zone.

LINE FIRING: Setting fire to only the border fuel immediately adjacent to a control line.

LOCAL WINDS: Winds, which over a small area, might differ from those appropriate to the general pressure distribution.

*LONG RANGE SPOTTING: Large glowing firebrands that can be carried high into the convection column, then fall out downwind beyond the main fire. Such spotting can easily occur one-quarter mile or more ahead of the main fire.

*MICROCLIMATE: A small site or habitat with essentially uniform climate, fuel characteristics, and burning conditions.

MID SLOPE: The area between the bottom of a canyon or ravine and the highest point of a ridge.

PARALLEL ATTACK:	A method of constructing a fairly straight fire line and burning out as the line is built.
RATE OF SPREAD:	A relative activity of fire in extending its horizontal dimensions. Expressed in chains/hours of forward spread, chains/hour of perimeter increase, etc.
REBURN:	Subsequent burning of an area in which the fire has previously burned but has left flammable fuel.
RED FLAG WARNING:	A term used by fire-weather forecasters to call attention to weather of particular importance to fire behavior.
RELATIVE HUMIDITY:	The ratio of the actual amount of water vapor in the air to the possible amount the air could hold at that temperature.
*RIPARIAN CORRIDOR:	A natural drainage, stream, or river (dry or wet) with foliage.
RUNNING:	Behavior of a fire spreading rapidly with a well-defined head.
*SADDLE:	A low point between to higher point on a ridge.
SAFETY ZONE (ISLAND):	An area for escape in the event the line is outflanked or in case a spot fire causes fuels outside the control line to render the line unsafe. In firing operations, personnel progress so as to maintain a safety zone (island) close at hand, allowing the fuels inside the control line to be consumed before going ahead.
*SHORT RANGE SPOTTING:	A fire producing sparks and embers that are carried by surface winds to start new fires beyond the zone of direct ignition by the main fire. The range of such spotting is usually less than one quarter of a mile.

*SKUNKING AROUND:	A fire that is not actively burning, except for an occasional bush or small spot.
SLASH:	Branches, bark, tops, chucks, cull logs, uprooted stumps, and broken or uprooted trees left on the ground after logging; also debris resulting from thinning, wind, or fire.
SLOPE:	The natural incline of the ground, usually measured in percent of rise (vertical rise divided by horizontal distance).
SMOLDERING:	Behavior of a fire burning without flame and barely spreading.
SNAG:	A standing dead tree or part of a dead tree from which at least the leaves and smaller branches have fallen. Often called stub, if less than 20 feet tall. Also called a widow maker.
SQUALL LINE:	A non-frontal line; usually a narrow band of thunderstorms with strong gusty winds extending across the horizon.
*STRIP FIRING:	Setting fire too more than one strip of fuel and providing for the strips to burn together. Frequently done when backfiring against the wind to create drafts that pull flames and sparks away from the control line.
SUBSIDENCE:	A descending motion of air in the atmosphere; of particular importance due to the heating and drying of the air as it contracts.
*SUNDOWNER:	A wind that can be very strong that occurs between 4 P.M. and 8 P.M.
SURFACE FIRE:	A fire that burns surface litter, debris, and small vegetation.

TOE	(SEE ANCHOR POINT)
THERMAL BELT:	An area of mountainous slope that typically experiences the least variation of temperatures, has the highest average temperatures, and thus the lowest average relative humidity.
TINDER:	Low density duff, peat, and rotten wood.
TOPOGRAPHY:	The configuration of the earth's surface including its relief and the position of its natural and manmade features.
TORCHING:	(SEE CROWNING OUT)
VORTEX TURBULENCE:	A sheet of turbulent air that is left in the wake of all aircraft.
WILD FIRE:	An unplanned fire usually requiring suppression action, or a free burning fire unaffected by fire suppression measures.
*WIND SHEAR:	Caused by two winds traveling in different directions. Causes erratic changes in wind direction and fire whirls.
WURST:	Wildland/Urban/Rural Structural Triage.

ANATOMY OF A WILDLAND FIRE

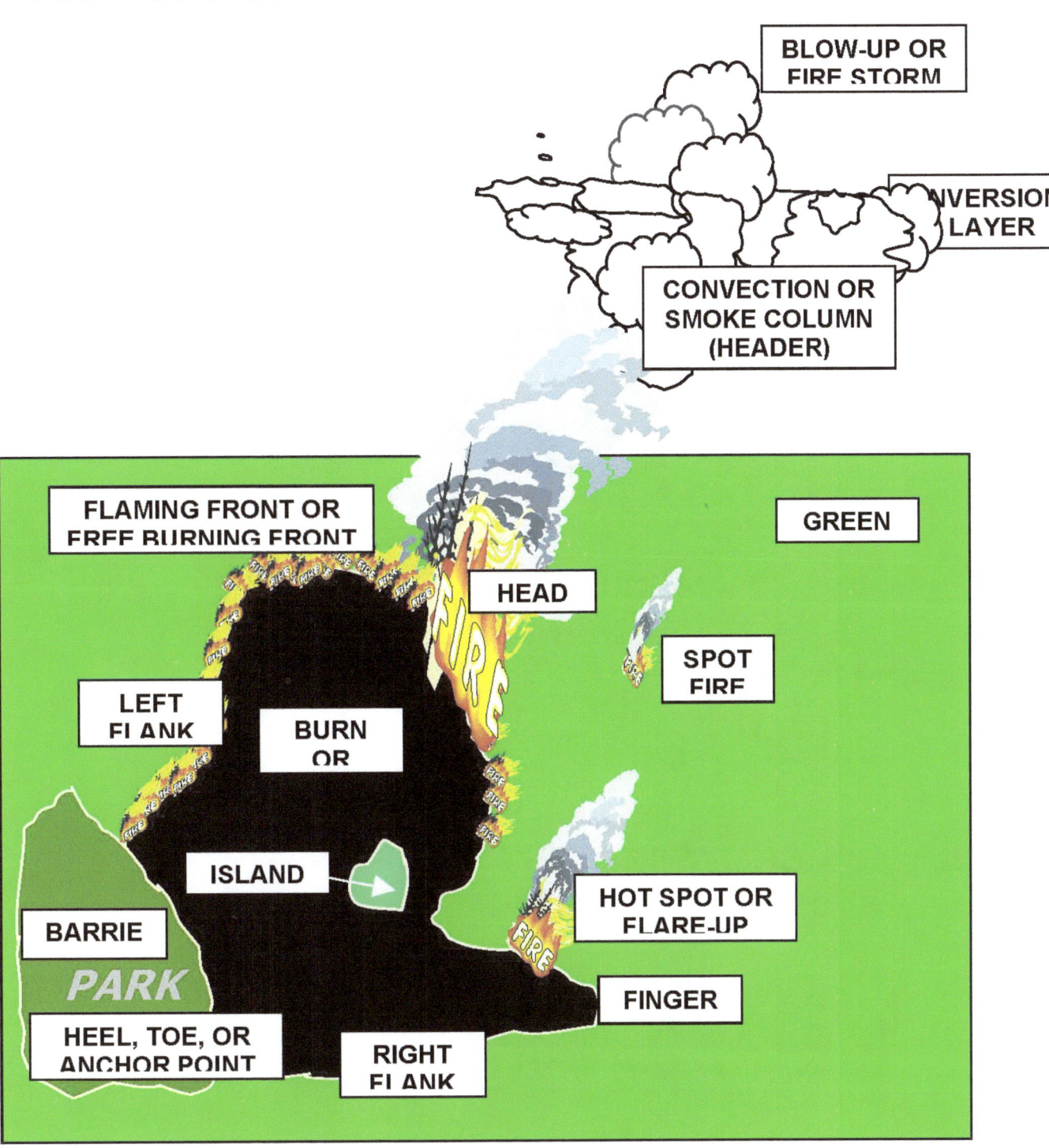

ANATOMY OF FIRE IN FUEL

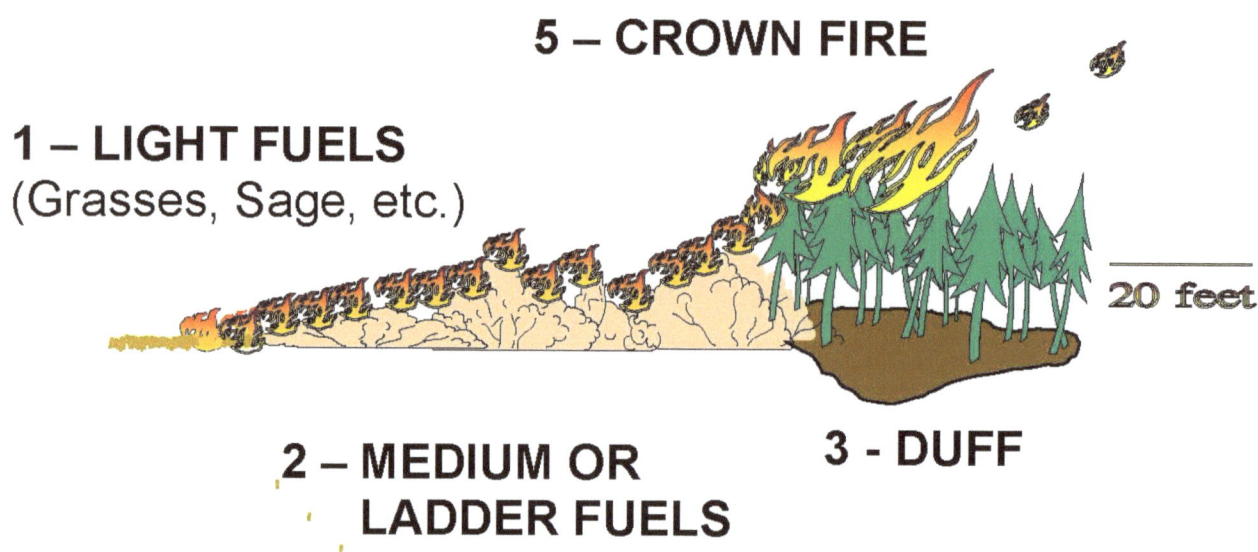

6 – FIRE BRANDS

5 – CROWN FIRE

1 – LIGHT FUELS
(Grasses, Sage, etc.)

20 feet

2 – MEDIUM OR LADDER FUELS

3 - DUFF

8 – CROWNING OUT
(Torch, Roman Candle)

4 – HEAVY FUELS OR AERIAL CANOPY

20 feet

7 – GROUND FIRE

ANATOMY OF WIND

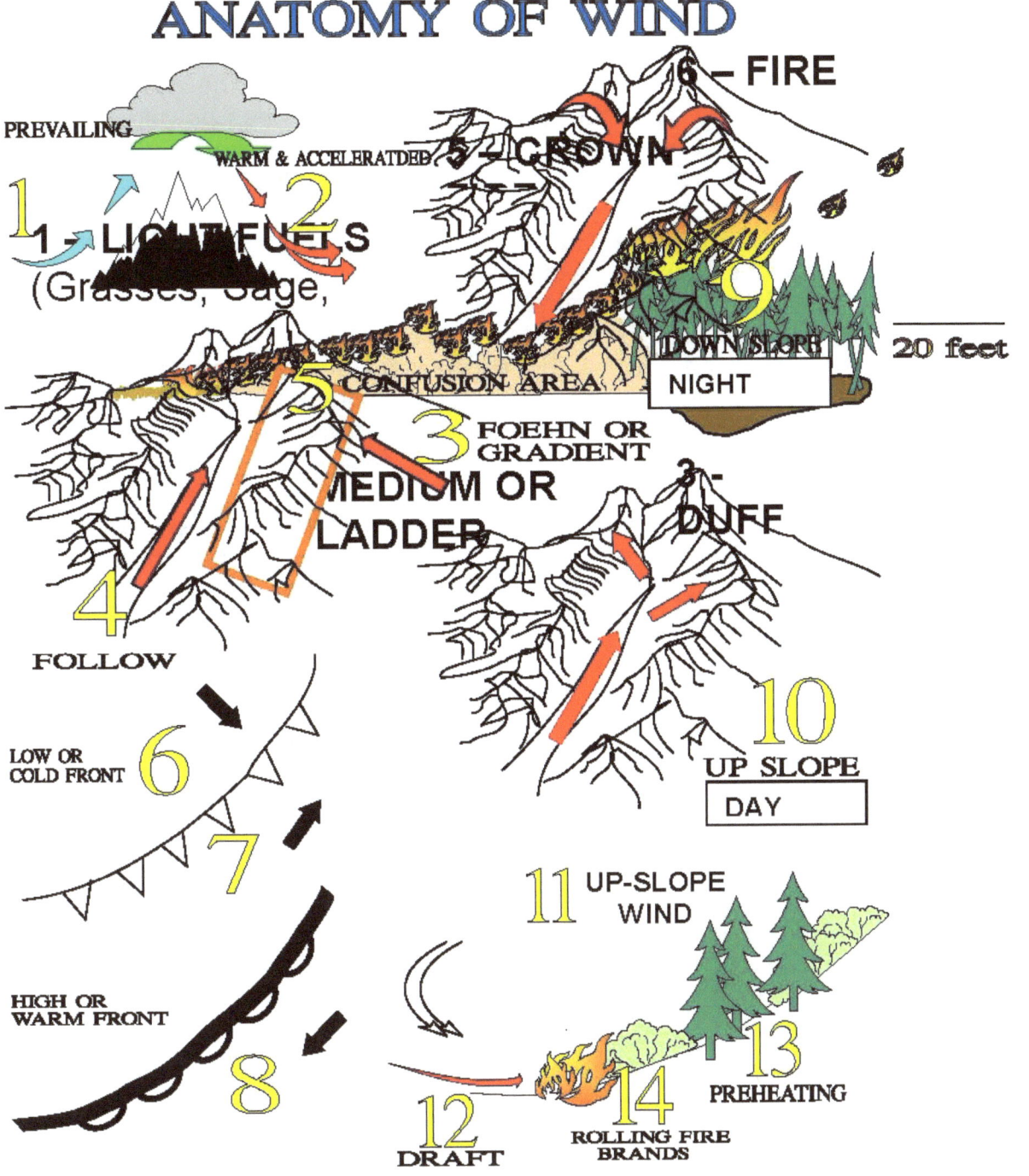

ANATOMY OF WIND

PREVAILING

WARM & ACCELERATDED

1 - LIGHT FUELS
(Grasses, Sage,

2

6 – FIRE

5 – CROWN

9

DOWN SLOPE   20 feet

NIGHT

CONFUSION AREA

5

3 FOEHN OR GRADIENT

MEDIUM OR LADDER

3 - DUFF

4

FOLLOW

10

UP SLOPE

DAY

LOW OR COLD FRONT

6

7

11 UP-SLOPE WIND

HIGH OR WARM FRONT

8

12 DRAFT

14 ROLLING FIRE BRANDS

13 PREHEATING

SITUATION ONE – NARRATIVE

Cedarville Engine One (a 250 GPM unit with 500 gallons of water) with two (2) personnel is dispatched along with two engines and a water tender from the Fire District to a fire along Highway 1 just outside of the City limits. The weather is hot and dry. Relative humidity is about 15%. Dry winds have been bowing at about 20-30 MPH for about two days. There has been very little, if any, measurable rainfall in the past three months.

Engine one is the first to arrive on the scene. The officer reports fire on both sides of the road with exposures up the hill in danger and requests direction from the incoming Fire District units. The officer in charge of the Fire District orders the Cedarville engine to protect the houses on Cedar Loop. He is about five (5) minutes out. When he arrives, he requests a third alarm. The next engine is ordered to attack the left flank of the fire to stop the spread toward the houses on Cedar Loop. The water tender is directed to support the house fire effort. The wildland fire is clearly out of control.

Eventually containment is gained, but not before 12 homes are leveled. Three homes are lost on Cedar Loop and nine as the fire entered the City limits, and 25 additional homes are damaged. In addition, 1,200 acres are burned.

What are the problems?

SITUATION ONE

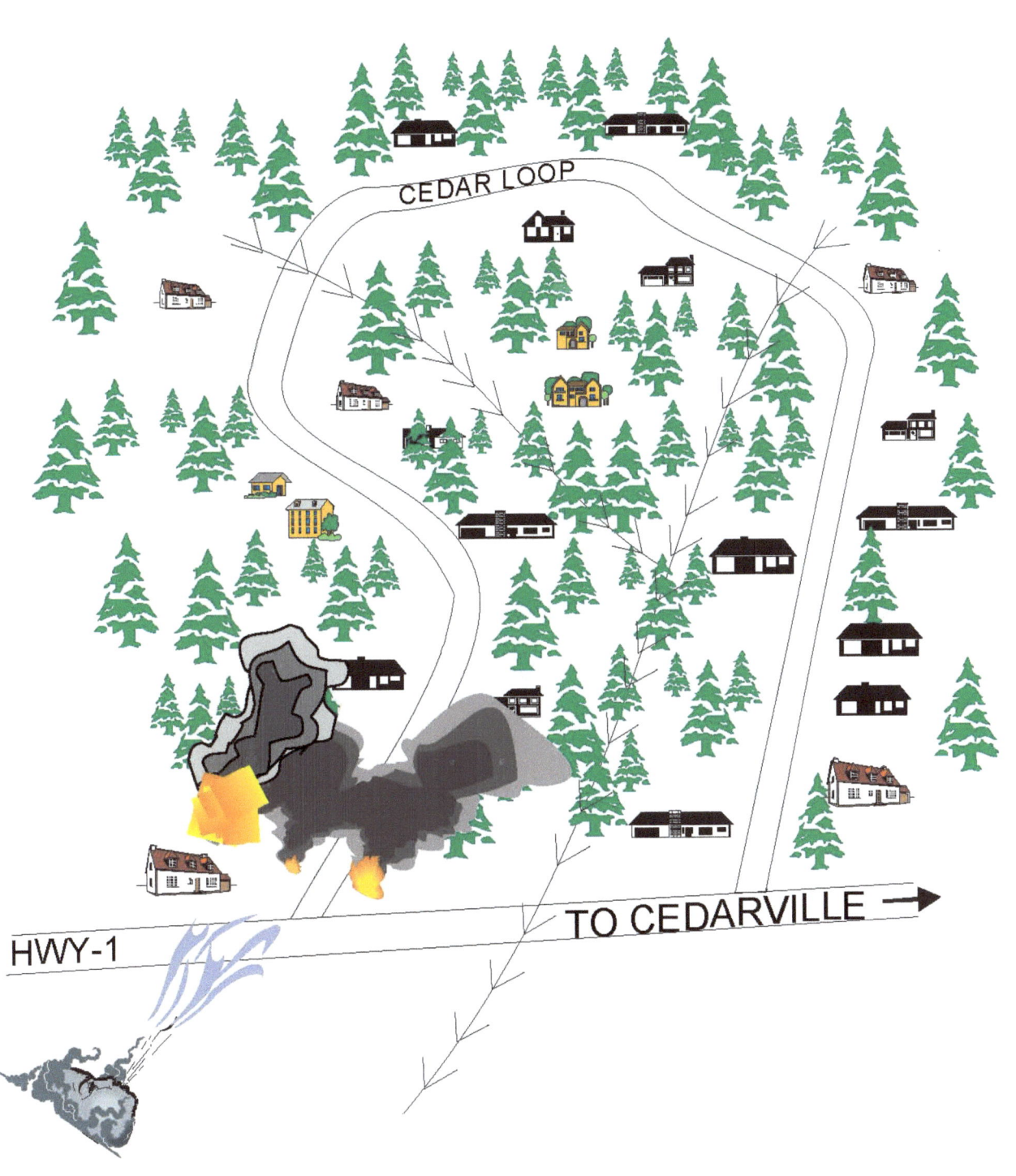

SITUATION TWO - NARRATIVE

The weather is hot and dry. Relative humidity is less than 10%. Dry winds have been bowing at about 20 MPH for past 24-hours. There has been very little, if any, measurable rainfall in the past four months.

A structure fire is reported at the bottom of the grade in an exclusive housing development. Three engines, a truck, and a chief officer are dispatched. The first arriving engine reports a working fire with spotting in the wildland area to the rear of the house. The dispatch center immediately starts a wildland response of a brush patrol and two wildland engines. The engine lays forward from the fire hydrant and attacks the house fire. Upon the chief's arrival, she orders two additional strike teams. Engine two is ordered to protect the next structure up the road. The truck is assists the first engine. All other units are directed to take-up defensive positions to protect additional structures.

The fire eventually consumes 2500 acres, eighteen (18) very expensive homes are lost, and numerous others are damaged, in addition one-person dies.

What are the problems?

SITUATION TWO

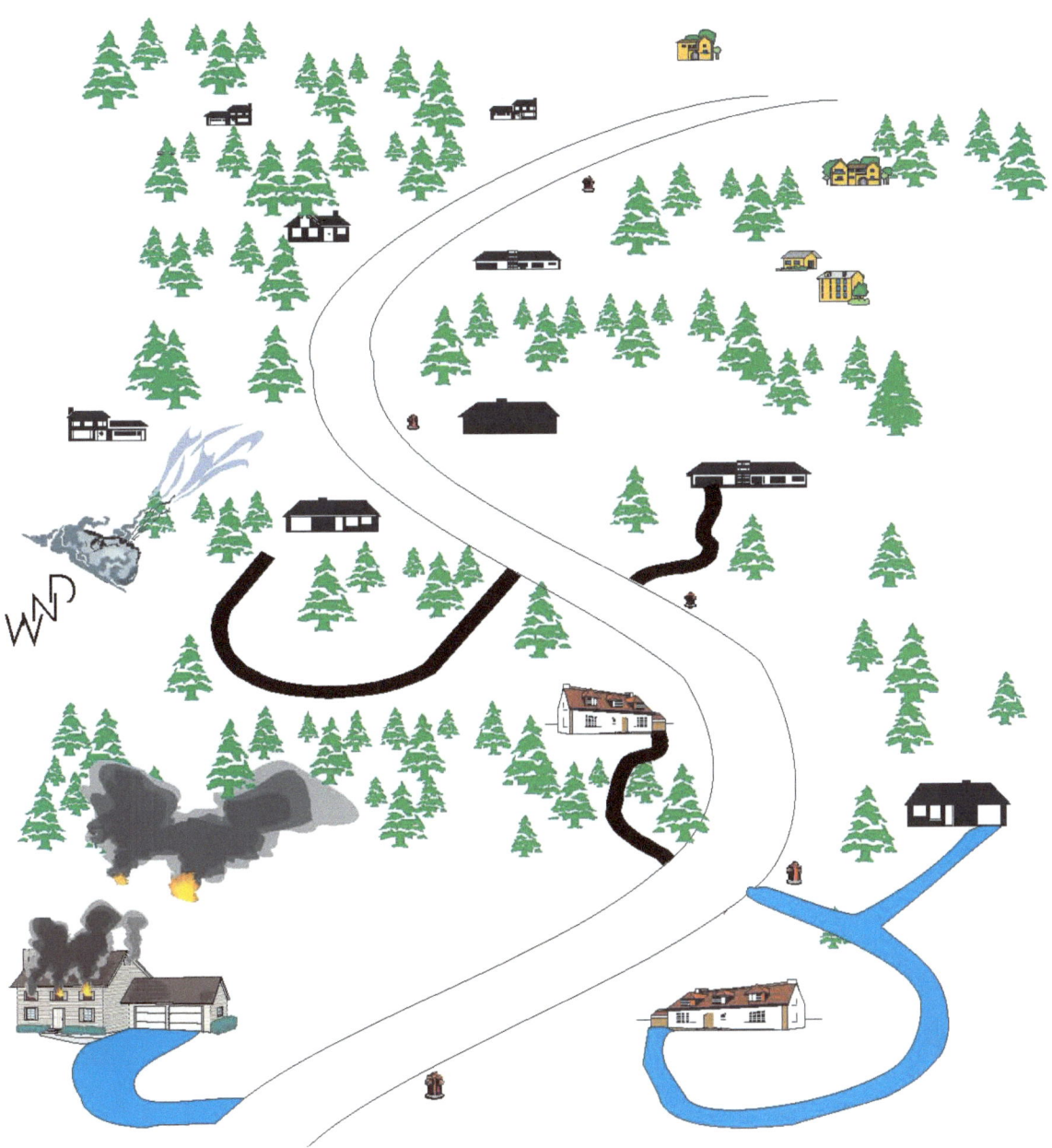

SITUATION THREE - NARRATIVE

It is the latter part of September and there has not been any rainfall for the past 60 days. The dry wind that has been blowing for the past two days has changed to a prevailing wind in the past six hours. Temperatures are unseasonably hot for this time of the year. At approximately 4:30 P.M., the incoming emergency lines at the dispatch center all light up at about the same time. People are reporting a large fire on the hill above a residential housing tract.

Access and water supplies in this area are very good. Two engines, a brush patrol, and a chief officer are dispatched. The first due unit reports seeing a header upon leaving the station. Two additional alarms are requested. Incoming apparatus is committed to structure protection on the uphill side of the fire and in the residential areas downwind. About 6 P.M., the wind is decreasing in velocity to almost a calm. At 6:20 P.M., the wind direction has changed 180 degrees and has increased in velocity to 25-40 MPH. A housing tract and the main part of town is now threatened. Three strike teams are ordered, but before they can arrive the fire races downhill and slams into the residential community and outskirts of town. The loss is 53 homes and four businesses.

What are the problems?

SITUATION THREE

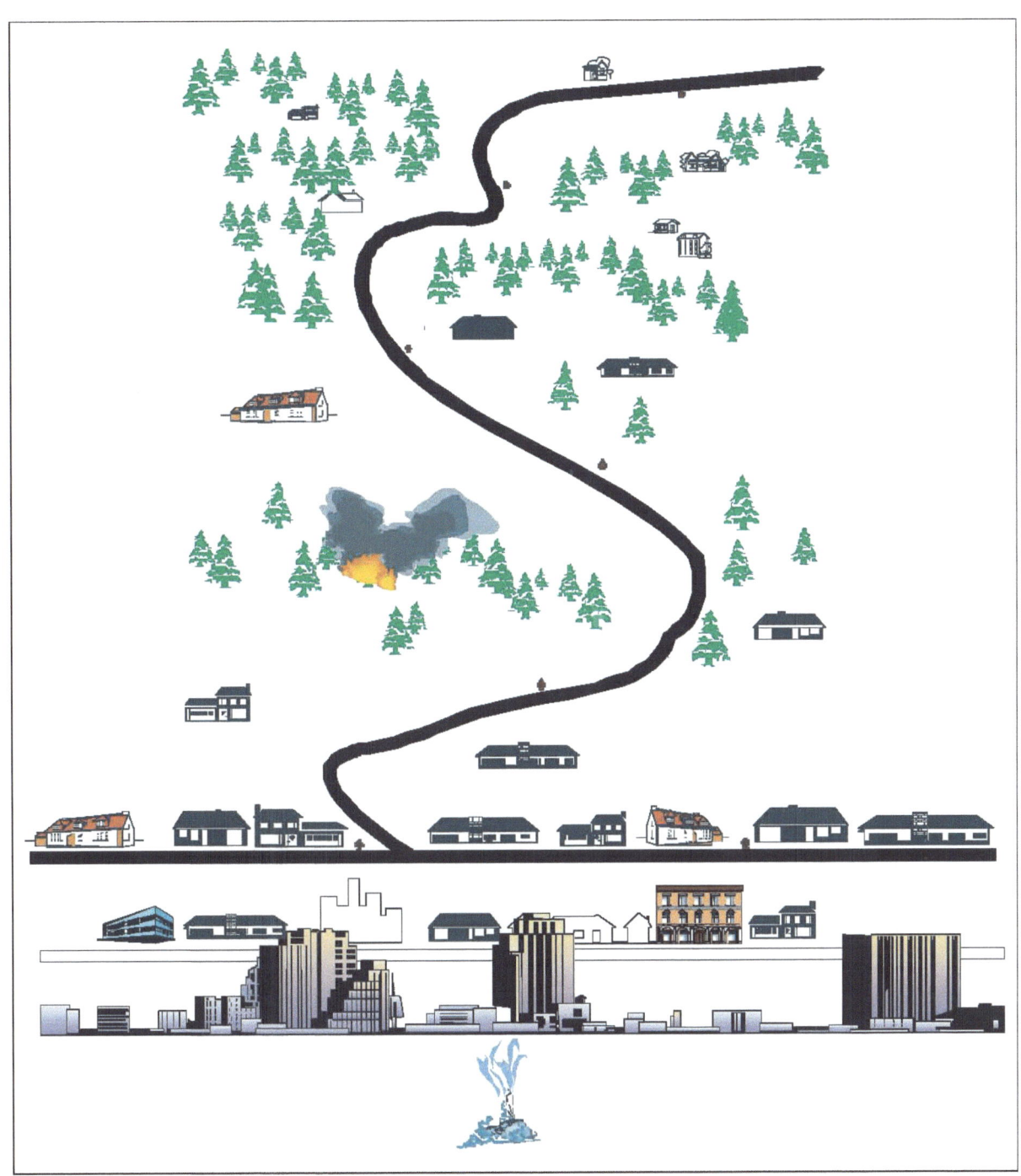

SITUATION FOUR - NARRATIVE

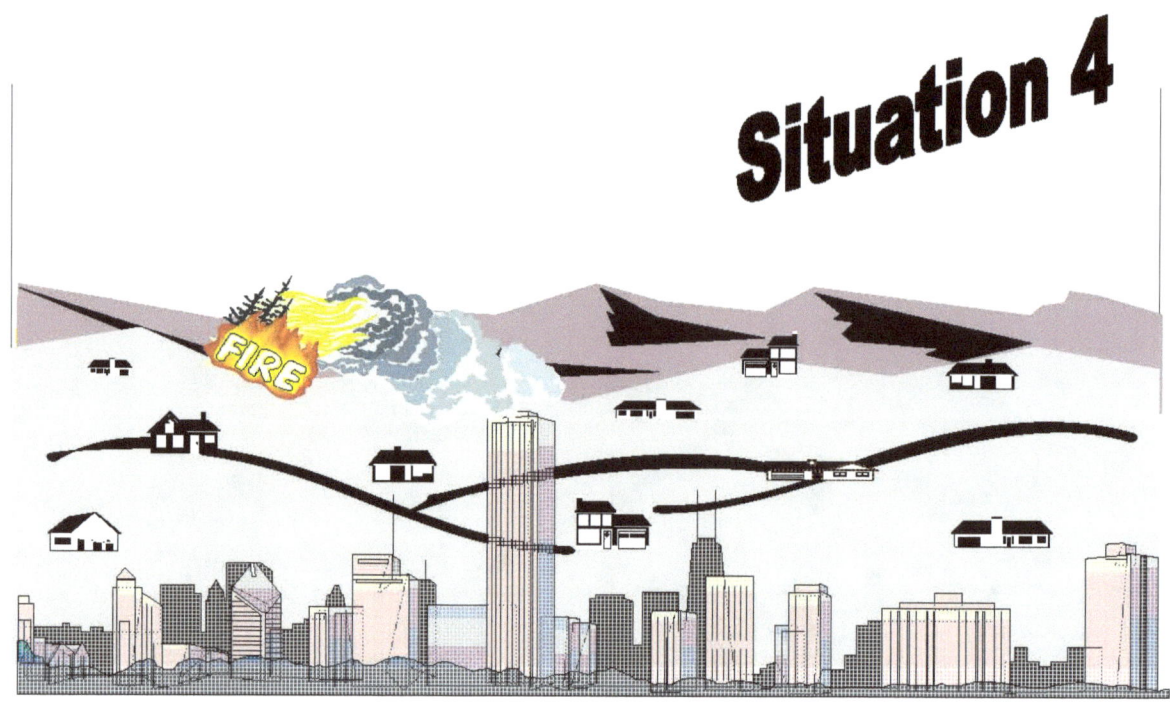

It is the latter part of October and there has not been any rainfall during the past 90 days. A fire that was contained in the early evening the day before has had apparatus and personnel watching it all night. About sunup a dry wind starts blowing. In the next 30 minutes the wind speed increases to more than 40 miles per hour. Crews hustle to get water on to some hot spots, but despite their efforts fire brands are pushed into the green and the fire quickly develops a large head.

Additional apparatus is requested, including two additional alarms and two strike teams. All units are committed to structure protection. Vision is poor from the ground and air support is requested. The first aircraft in the area reports a major fire front spreading in two directions. Numerous homes can be seen burning.

Evacuations are ordered. In the ensuing chaos, one firefighter, a police officer, and six residents are killed attempting to evacuate. Numerous firefighters, police, and civilians are injured. More than 400 homes are lost.

WHAT ARE THE PROBLEMS?

SITUATION FIVE - NARRATIVE

The weather is hot and dry. Relative humidity is 11-13%. Dry winds have been blowing at about 10-20 MPH for the past 24 hours. There has been very little, if any, measurable rainfall in the past four months. A critical weather pattern has existed for some tune.

The fire is burning on one side of the highway with a gently up slope. Fuel in this area is medium with a mix of sage and 12-18 inch grass. Homes are located on both sides of the road. A spot fire occurs in the grass on the other side of the highway. Two engines companies attack the spot and are gaining the upper hand leading to control.

A second spot occurs a short distance from the first spot and as personnel are stretching more hose a fire whirl develops on the other side of the road. It quickly builds to 200 feet high and approximately 100 feet wide. Pushed by winds reported in excess of 100 M.P.H. the fire whirl moves across the road and over runs the firefighters working on the spot fires resulting in 4 personnel being seriously burned, one is critical.

What are the problems?

WHO IS AT FAULT?

SITUATION 5
EAGLE FIRE

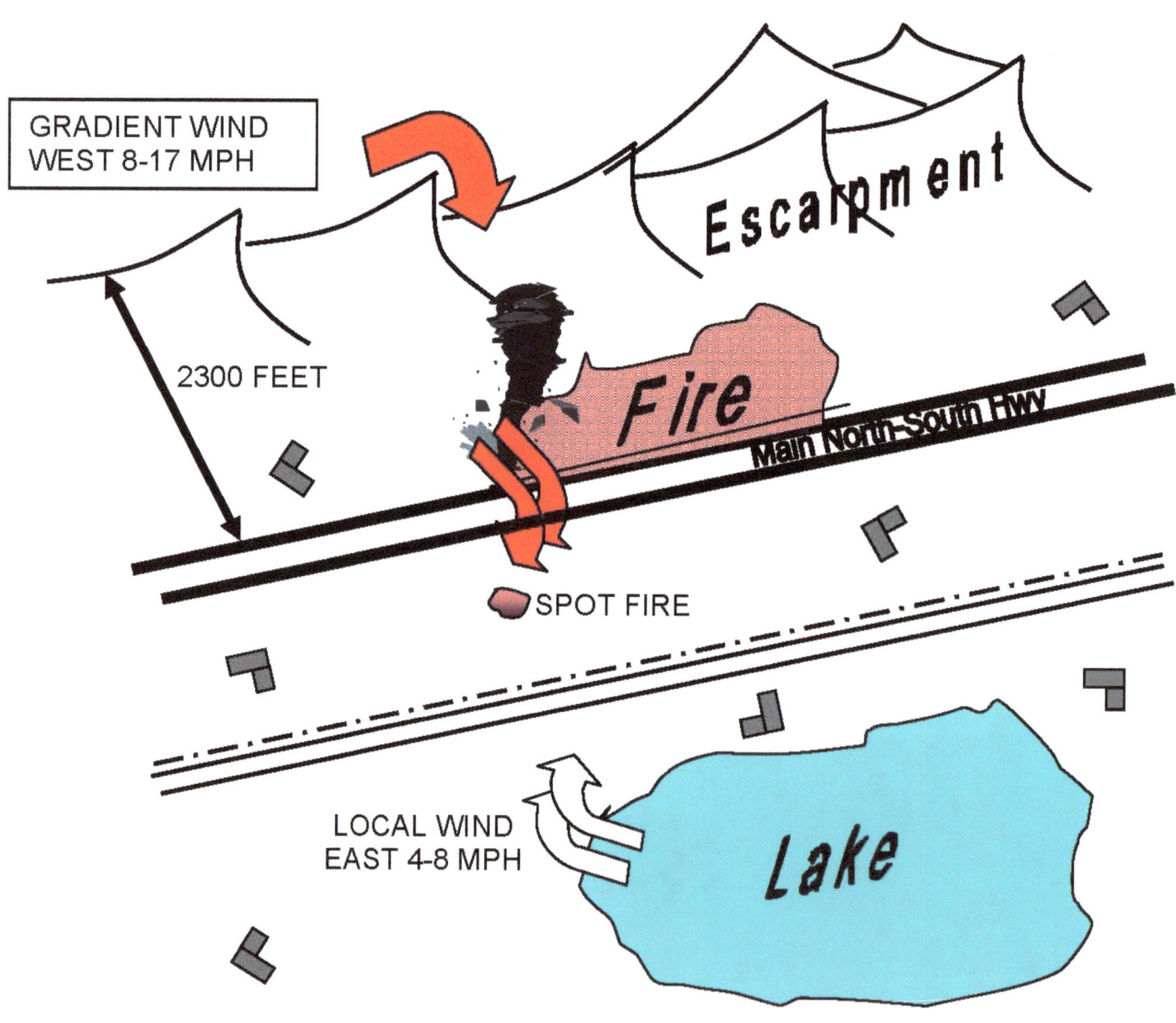

GRADIENT WIND
WEST 8-17 MPH

Escarpment

2300 FEET

Fire

Main North-South Hwy

SPOT FIRE

LOCAL WIND
EAST 4-8 MPH

Lake

SITUATION SIX - NARRATIVE

On September 19 the weather is hot and dry. Relative humidity has been less than 20% for the past week and a half. There has been very little, if any, measurable rainfall in the past four months.

A fire is burning in an area of grass and light brush. The fire is burning on a grade, and has slopped over the road and is burning on both sides. A comp any coming down the grade is assigned to pick up the slop-over. The officer spots his engine on the road above the fire, and he and the other two firefighters proceed to lay a hose line down. On a return trip to the engine to pick up more hose one firefighter notices a flare-up, and after two attempts contacts the other two personnel by yelling. The two firefighters start to walk out carrying unused hose with them. The fire has paralleled their position and is now making a run at them from two directions. They attempt to make it to the road up a steep bank. One firefighter does not make it; the other receives serious burns.

How Many of the Ten (1O) Firefighting Orders Were Not Followed?

SITUATION 6
MACK #2 FIRE

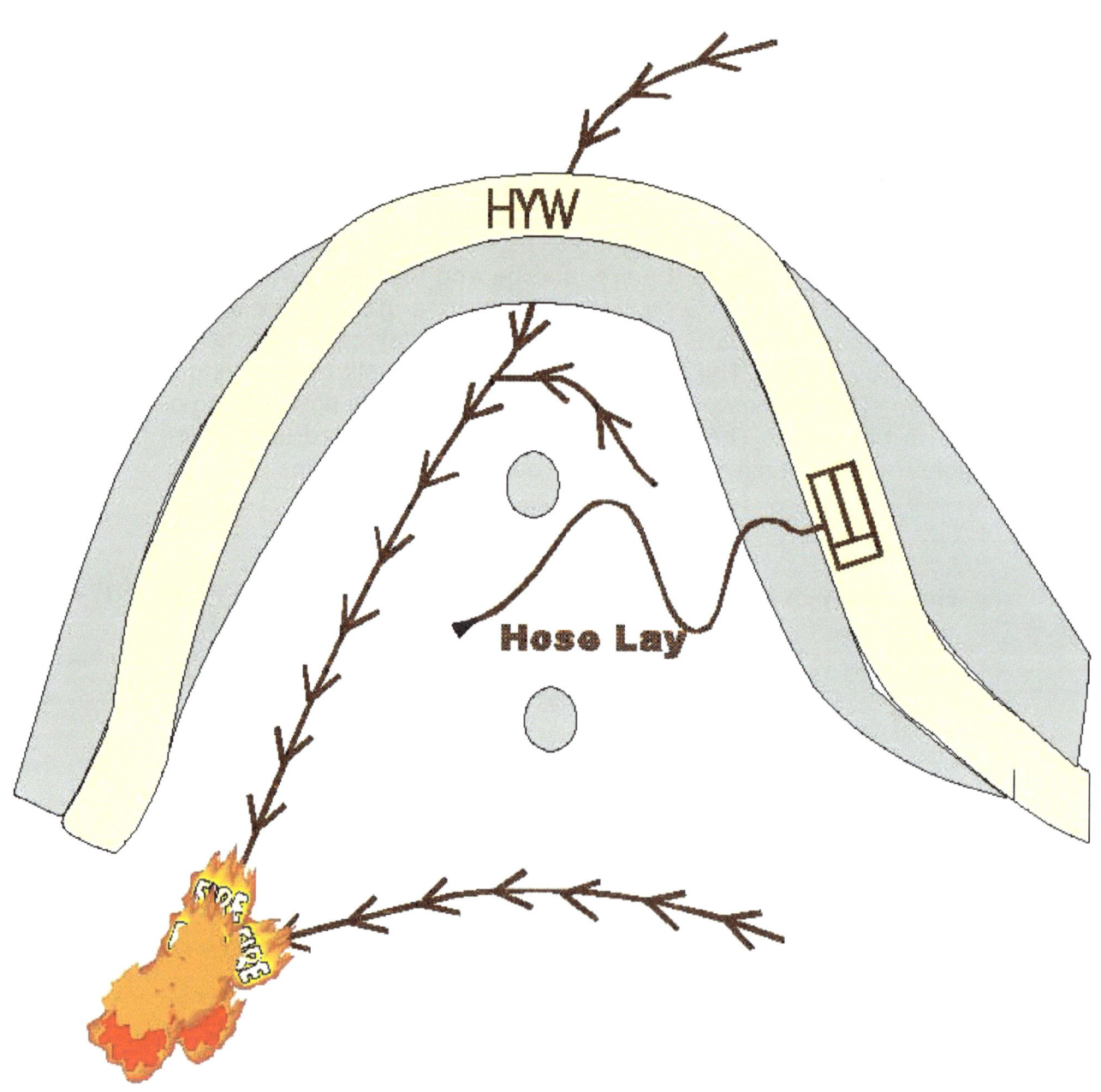

SITUATION SEVEN - NARRATIVE

It is mid-October, the weather is hot and dry. Relative humidity has been less than 20% for the past week and a half. There has been very little, if any, measurable rainfall in the past three months.

A fire is burning in a region with 20-year-old brush growth. The previous winter and spring were fairly wet which allowed for a heavier-than-normal growth of grass. The fire has consumed approximately 1000 acres and is spotting 1/4- to 1/2-mile ahead of the main fire. The area of the fire has numerous exclusive home 5. A chief officer orders two engines each with four personnel to protect homes ahead of the fire advance. While proceeding to their assigned areas the engines are caught on a saddle with chimneys leading into it. The fire is approaching the saddle from both sides.

One engine parks next to a cut bank, and the firefighters use the apparatus and the bank to shield themselves as the fire overruns their position. The other company abandons their unit and runs up the center of the road. All firefighters received burns.

The fire eventually covered 4,500 acres, 80 very expensive homes are lost, and numerous others are damaged. One firefighter received third degree steam burns.

What are the problems?

SITUATION EIGHT - NARRATIVE

Your engine was directed to protect structures in a dead end canyon on a fairly slow moving fire. Once in the canyon your division supervisor advises you of a major blow-up that will compromise your safety and you are ordered to evacuate.

You start out leading two residents in a pick-up truck. As you proceed up the road your path becomes blocked by the fire's advance. Spot fires are occurring all around you and a wall of fire is rapidly approaching. Your engine and the pick-up are going to be overrun in a mater of minutes. Given the choice between the rocky area devoid of vegetation, road, river, vehicles, or riverside sandbar, which place, would you use to deploy your fire shelter?

Prioritize up to six locations your group feels would be safe to deploy your fire shelters. Site number one would be most preferred and site number six the least preferred. Be ready to explain why you prioritized the locations as you did.

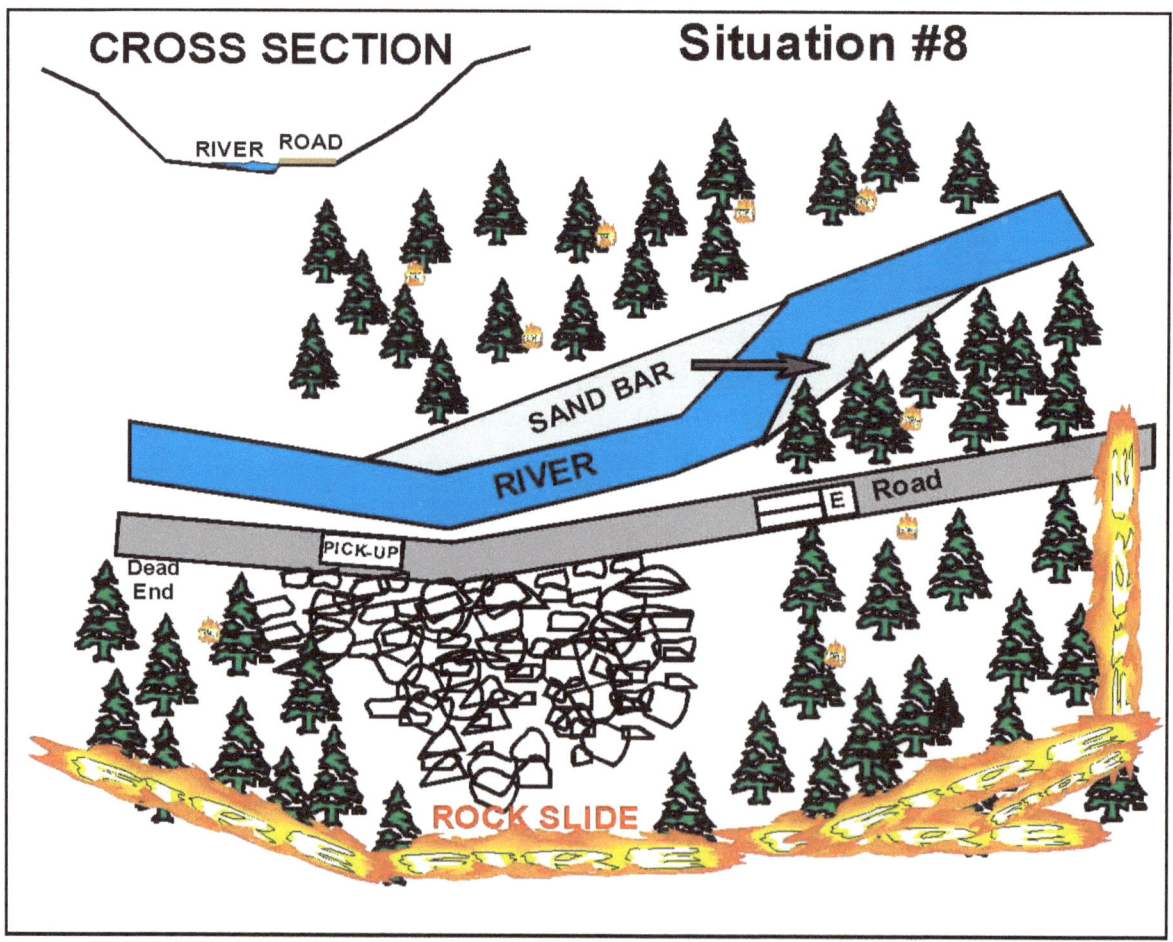

INFORMATION SHEET

WILDLAND/URBAN/RURAL STRUCTURAL TRIAGE (WURST)

INTRODUCTION:

The fastest growing fire problem in the United States is the potential of a Wildland/Urban Interface Fire. Communities are no longer concentrated in an urban core. As our population continues to grow, homes are being built in what once was uninhabited wildland. The problem is not isolated to the western United States. Each year more fire departments and officers are finding themselves faced with a wildland fire with the potential loss of structures.

INFORMATION:

Community planning, public education and hazard reduction programs ultimately may decide the answers to the wildland/urban interface fire problem. Even with these elements of a fire defense system, fire officers are still going to face some tough decisions when a fire starts in the wildland/urban interface. When these situations occur, the best defense may be a strong offense. In other words, responding companies should be taught to attack aggressively these fire problems to control the spread of the wildland fire. Aggressive attack often reduces the potential loss of structures. Yet, at some point we may be faced with a wildland/urban interface fire that is beyond the control of responding resources. When this occurs officers have to switch tracks and move to defending homes.

We refer to homes specifically because during the wildland/urban fire outbuildings such as barns, garages, sheds, etc., often become secondary. Generally, structures of this type do not receive protection consideration unless they are the only structure threatened or are close to a house. (It is a simple fact that some houses and auxiliary structures must be left to stand alone without protection. Making this decision will be one of the most difficult a fire officer will ever have to make. How do you know what houses to protect if you cannot protect them all?) What companies and officers perform in this situation is what I call Wildland/Urban/Rural Structural Triage (WURST). They have to quickly sort through several factors about a home to decide if some type of action is warranted or not. To help in this process I have put together the WURST Model. Just like emergency medical triage involving human life when personnel have to become selective as to whom to help, so it is with structural triage. However, what is decided here is what to leave to the whim of Mother Nature.

The is how the WURST process works. Take seve5ral homes in wildland/urban interface fire area where control of the wildland fire is not likely. The first question is: Is the home burning or not?

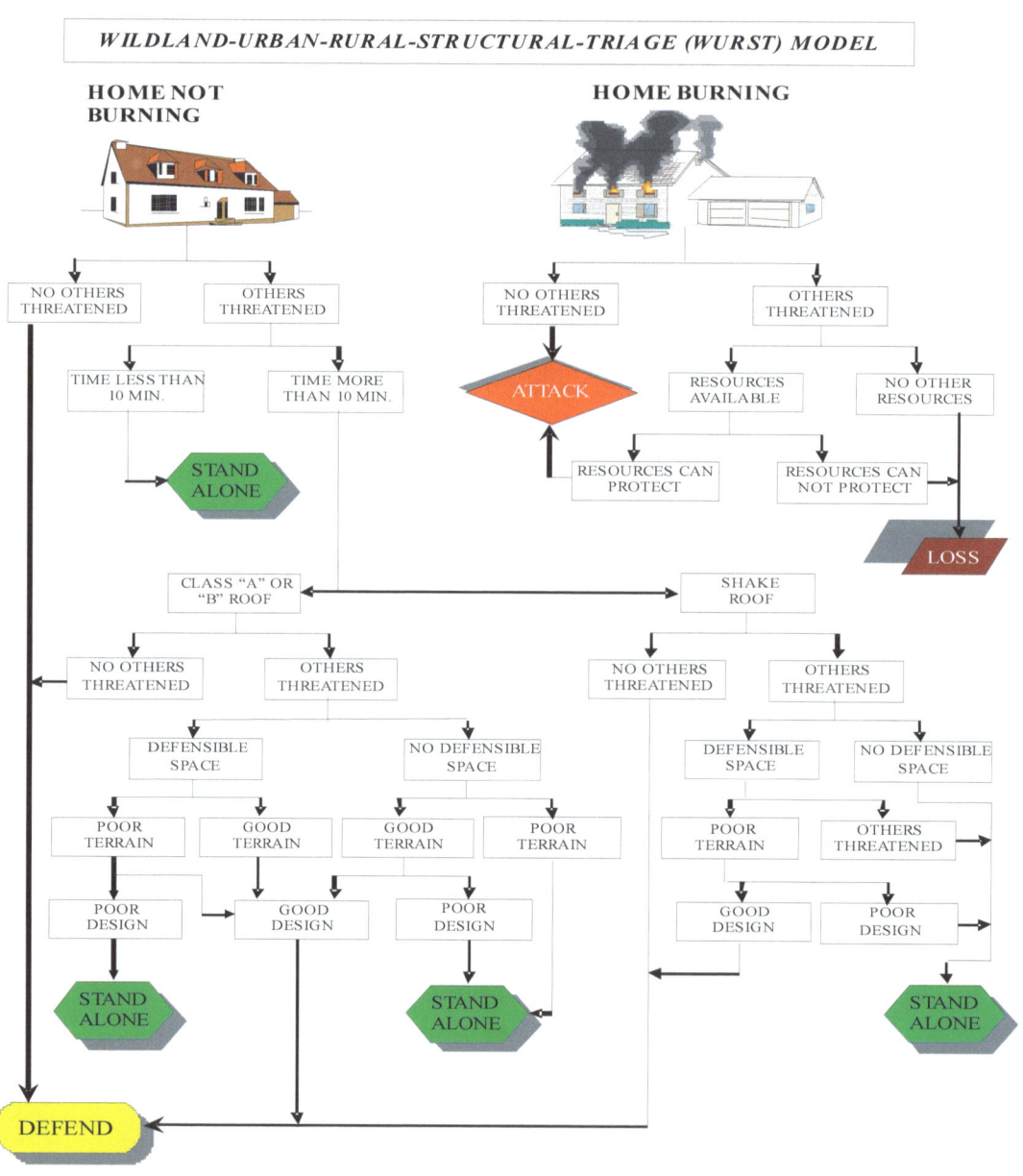

If the house is burning the decision to take suppression action becomes a matter of available resources. In many of these situations you are working by yourself. If this is the case, burning structures must be written-off to save others that are not yet on fire. If other resources are available or no other homes are threatened then fire attack may be advised. Yet if these other resources cannot protect the other exposed homes and also provide you with assistance, then the burning structure needs to be moved to the loss column.

Most of the time when homes are not burning, you need to be at least ten minutes ahead of the advancing fire front. This time allows a company to get set up to defend a home or other structure. Usually, at best, the availability of resources is limited to one unit to each exposure. Sometimes one engine can save several homes that are close to each other. Cul-de-sacs are often good locations for one engine to protect several homes. If there is a threat to other homes you may have to leave the house to stand alone, especially if there is less than ten minutes.

When a company has ten minutes or more, the decision to defend the structure may rest on the type of roof. Shake roofs are difficult to protect from flying fire brands and should receive less consideration than composition or better roofs. Homes with shake roofs can be saved if they have defensible space around the structures. Defensible space should be a minimum thirty feet cleared area on all sides of the structure. On steep up-slopes the defensible space may have to be up to two hundred feet depending on fuel type and its volatility. A good rule of thumb for the amount of defensible space based on slope is fifty feet for each 10% slope. If trees overhang any roof or if a concentration of litter is found on the surface defending the structure will be difficult and most likely, impossible, if the roof is wood shake.

If there is defensible space around a house, then terrain becomes the next consideration in the WURST model. Good terrain is a fairly level surface. Homes on slopes of more than 30% may be very difficult to defend. Houses that sit in saddles, at the top of steep slopes, on ridges, and at the top of ravines or chutes are not good risks. These homes may have to be left to stand alone so resource can protect other houses in more advantageous positions.

Structure design also must receive consideration when selecting a structure to defend. Houses with exterior wood decks that are not enclosed tend to trap hot gases and burning embers blown by the wind. Buildings that do not have non-combustible foundations or well enclosed foundation skirts pose more difficult protection problems. Buildings designed with defection protection (see next page) that is a third higher than the fuel below have been known to stand alone with no intervention while other homes around them without such protection burned to the ground. The WURST system is built on the 25% rule for triaging structures. Each of the four defense elements: roof type, defensible space, terrain, and design is assigned 25%. A house with 50% deficiency is considered marginal or a poor risk if there are other homes that are in danger. Some individuals weigh a shake roof at more than 25% deficiency. Yet, a shake roof by itself is not a death sentence for a structure. I have seen shake roof buildings (with no other deficiencies) defended successfully with as little equipment as a garden hose.

TWO TYPES OF DEFECTION PROTECTION

When triaging structures, companies need to be trained to work from water carried on the units and on-site resources such as garden hoses, swimming pools, ponds, or streams. Most water systems are likely to fail and should not be relied on too heavily. Systems often fall victim to loss of power to pumps and overwhelming demand from elevated tanks. Master stream appliances from units may work well in the lowland areas, but they will greatly reduce or make pressure useless in the higher water demand zones. Generally, higher zones are located on the ridges where the need for water is the greatest and most of the structure loss will occur without an adequate water supply. Firefighting apparatus operators should be trained to refill or top-of the tank at every opportunity.

The purpose of the WURST model is to help you identify structures that can be defended easily. Yet, we must not fail to recognize that taking the easy building to defend may not be the right choice either. A building located on flat ground with a non-combustible roof and large defensible space most likely will stand on its own. In this case select another structure that have one or more problems as identified through WURST to maximize the use of your resources.

COMMUNICATIONS

INTRODUCTION:

Communication is the method by which the plan is conveyed, the organization structure identified, and the controls enacted. A manager can plan, organize, and design controls with poor or ineffective results if he/she can not adequately communicate. Communications is a word that means different things to a wide cross section of the emergency community. Webster defines communication as; the act of transmitting; a giving or exchanging of information, signals, or messages by talk, gestures, writing, etc.; but to most of the emergency personnel it is the transmitting of information by use of the radio.

To the emergency manager communications has to be much more than the radio, that which transpires between the incident commander; supervisors, and subordinates through communications has to be clearly defined, sufficiently discussed and acted upon decisively if performance is to measure up to expectations. Emergency communications is performance oriented. The power at the command of managers, supervisors, and subordinates cannot be released; directed or coordinated without a network that can and does communicate vital and individually meaningful information.

INFORMATION:

Communications has been called a science, an art or a skill. It does not matter to which vehicle you feel communications belongs so long as you recognize that emergency communications is a leaned medium and what we do from a communications standpoint will have a major impact on all incidents.

How many times have we in the emergency field heard the words, "It was a failure to communicate" or "a communications breakdown" as the underlying reason for an error? It seems that each time we objectively conduct a post incident analysis, communications or the failure thereof can be traced to well over *50%* of the problems that were encountered as the operation progressed. The reason for this is that the ability to communicate is too often taken for granted especially during the first minutes of commitment.

There are approximately 600,000 words in the English language. The average educated adult uses about 2,000 words in daily usage and the 500 most frequently used have 14,000 meanings. This points out an important thought to remember words do not have meanings how people interpret and translate them into action does. The radio, although an excellent tool, is more often than not abused or misused causing problems.

Problems in Communicating

Problems in communication arise because we forget that our individual experiences are never the same as the people with whom we work. Managers with more in-depth knowledge will inadvertently omit key elements in an order or directive, leading to misunderstandings. This occurs because usually the wrong mode or method of communicating is chosen. Remember no two people ever hear or for that matter see things in exactly the same way.

Perception or what we see is often mixed with oral communication from differing locations on an emergency causing breakdowns. Problems or situations viewed on opposite sides for example often present remarkably similar circumstances with little or no interconnection.

People communicate in a variety of ways. They do this by physical touch, by visible movements of portions of their bodies, and by symbols either spoken or visible. Failure in communications is often due to a misunderstanding of symbols or inadequate transfer of perception of these symbols. During emergencies the loss of any one of the following elements of communication is most likely to lead to communications breakdown: Spoken symbols, visible symbols or visible body movement. Note:
Two of the three elements are visible; therefore, it is reasonable to expect that if you or your agency relies heavily on the radio that sooner or later a communications breakdown will occur.

Spoken Symbols

Spoken symbols have the most impact of all the communication elements. More misunderstanding occur because of what is said than found with any other element of communication. Spoken symbols during emergencies are those very important words that clarify and/or release authority and responsibility to another person. The simple stating of things like, "you are in change", "coordinate with", "report to", etc. can have a dramatic impact on the outcome of emergency operations.

Spoken symbols on emergencies can be said to erase doubt in the minds of subordinates. Often times these symbols are action verbs that explain what has to be accomplished.

Another use of spoken symbols is found when the radio is the only means of communicating. A spoken symbol in this case attempts to tie individuals together by referring to common knowledge shared by both. Examples are; past experiences than could be applied to the present situation, training methods, step by step directions, etc.

How we say a thing, the tone of our voice has an impact on personnel. Excitement on the part of an Incident Commander in communicating can cause violations in even basic safety precautions.

Visible Symbols

Visible symbols are simply what the manager and subordinate can both see. Maps or drawing used to clarify communications are excellent examples. The old saying " a picture is worth a 1,000 words' should not be overlooked when managing emergencies.

Visible Body Movement

Visible body movement is often referred to as non-verbal communication. Visible body movement is most often thought of as "pointing" to identify places of common reference. But, non-verbal communication is also the look in a person's eye, facial expressions and body position, all of which can effect the meaning of our words or the actions of personnel.

Watch people when you are communicating. If their eyes are glassy or very wide you may not be communicating. Are they twitchy? They may be uncomfortable with what you are saying. Can you see a look of confusion in their facial expression or is it disagreement? These are but a few examples there are others watch for them to become a successful emergency communicator.

Timing and Space and Their Communication Impact

Timing and space will affect communication. A person or subordinate must feel comfortable for communication to be effective.

Timing is important. Avoid when possible communicating with resources that are not on scene. They should be at the very least staged nearby. This will cut down on the need to change orders or correct an action that is not based on the current situation. Know when to communicate important information. Provide only what is necessary at the appropriate time.

Space has a large impact. The closer you are to a subordinate the easier it is to read the person or provide the correct type of communication. For example, during an emergency personnel can become very excited or apprehensive; a hand on the shoulder can have the desired calming effect. This is a type of communication that is not talked about in most material on emergency communication; the impact though can be dramatic.

METHODS OF COMMUNICATION ON EMERGENCIES

The problems of communication on emergencies comes down to two basic factors clarity and conciseness. If a communication is clear it is put together in a logical manner progressing form point to point. If it is concise, then it delivers the message of what is to be accomplished. Communication on emergencies can be handled through anyone or combination of five methods. They are radio, phones, runners, written messages, or face to face communications.

The Radio As A Means Of Communication

The radio is the most common, easily accessible, and convenient method of communicating. It is also the most abused, overworked, and disastrous tool ever invented when utilized incorrectly by the emergency manager. If these two statements appear to be in conflict, they are, just as the radio often is during an emergency. The more frequent communication is identified as a problem when analyzing past situations, the more prevalent the use of the radio.

Discipline is the answer to many of our radio communication problems along with recognition of when not to use the radio. Discipline involves what is known as protocol. Every agency, no matter how large or small, should have some written policies on what priorities are to be placed on certain radio traffic. Generally it should be noted that use of the radio is primarily designed to aid key supervisors in managing an emergency situation. Other personnel should not routinely utilize the radio unless contacted by a command manager.

In a situation involving an injury to personnel or potential injury agencies should adopt a radio notification system based on the Incident Command System (ICS) used by most wildland firefighting agencies. That communication signal is "EMERGENCY TRAFFIC" followed by a plain text description of the problem.

Supervisors need to recognize when to use the radio and when other forms of communication would be more effective. Generally those in charge should be aware that detailed communications such as those in a directive are not suited for the radio. Likewise assignments to key personnel need to be made on a face-to-face basis. The reasons for this should be apparent. The radio, although efficient, is not as effective as other method because it does not allow a free exchange of information from one person to another. The exception to this would be very simple communications involving one task.

The radio is an excellent tool for a manager to receive feedback and control activities, but as a sole means of managing communications on an incident, it is less than adequate. A major area of agreement concerning the radio is that it saves time. However; this also is a misnomer; when supervisors consider the number of error that are make when only the radio is used to communicate. Errors from strict use of the radio can cause longer delays, which in turn create confusion and more involvement from a time standpoint.

Phone Systems as A Means of Communication

The best means of transmitted communication is by a phone system. There are four basic types; they are public, cellular, satellite, and portable.

Public Phones

One of the best, most often over-looked, communication resource is the public telephone system. Public phone systems classification includes any equipment that may be use to call numbers outside of one area. This could include private phones owned exclusively by a private party as well as public utility systems. The only limiting factor in the use of these systems appears to be training and educating of emergency personnel as to their operation and advantages.

Training on different types of systems creates one of the largest problems. Home or single line installations are simple but business systems vary. Most business phone systems require some type of code or prefix before being able to dial outside numbers. This usually is a single number such as 8 or 9 but can be a coded number of3 or more digits. Some systems require one procedure during business hours and a different one after hours.

The use of the phone in command situations permits a much freer flow of information between the incident commander and the communications center than that which could transpire over the radio.

Caution should be urged and control exercised in regard to who can contract the communications center by phone during an incident. If some type of control is not exercised the communication center can become paralyzed. This is especially true on large incidents where they may have to use the phone to make outgoing calls to order resources. For this reason every communication center should have at least one line that is for outgoing call only. Special lines of this nature can be ordered from the local Phone Company.

There are ways to get more mileage out of the phone system. Many closed systems have what is known as conference dialing where one or more parties may be on the line at the same time. Likewise public systems often have what is known as a conference operator. This person can put you in contact with many parties at one time. The only criteria is that you must know all the phone numbers you desire to contact and how to contact the conference operator. Normally the operator can be reached by calling the normal operator and asking to establish a conference call.

Cellular Phones

Many agency staff vehicles are equipped with cellular phones that can signals from almost any location, except in some rural locations. This makes the use of the phone on emergencies become increasingly more important. The major limitation of cellular phones is their ability to cause gridlock in the phone system. On major emergencies so many people sometimes use the cellular phone that communication centers are often overwhelmed.

Satellite Phones

A satellite phone can prove to be invaluable in areas without cellular service or where such service is intermittent. The largest drawback to these phones is cost of both the phone and air time. Yet, in areas without other phone services this may prove to be a small obstacle.

Portable Phone Systems

To solve the problems of radio frequencies becoming over-loaded with excessive traffic, some agencies have designed and constructed or purchased self-contained portable phone systems. These systems are battery operated, battery assisted, or sound power. Their use can be extremely helpful in running a incident where large support function are set up in one location. All the systems operate in much the same manner. The only difference being the maintenance factor required on battery or battery assisted systems. The most preferred is the sound power type.

Runner as A Means of Communicating

Runners or aids do have their advantages and often times are more effective than the radio or phone. The way to maximize this resource is to avoid long complex orders and allow supervisors to feed back through the runner.

Long and complex orders are apt to be misconveyed when using a runner. Anyone who has played communications games will attest to the fact that something is invariably lost in any relay of information. Attempt to hold runner communications to the transmittal of no more than 3 new items or changes in strategy or a combination of both, especially with more complex the communications.

The best use of a runner is to select a trained manager or subordinate that thinks like you do. In this manner the runner can add that little touch that may be necessary to get the point across. Also a trained manager can more readily transmit feedback to you. This is because the runner has the ability to tie loose ends together through personal observation.

Written Messages as A Means of Communication

Although written messages are seldom used they can be an excellent communication tools. Diagrams, maps, sketches, etc. can provide more information than any other means of communication. Also written directions limited to one task delivered by a runner can be extremely useful.

Face-To-Face As A Means of Communicating

The most seldom used means of communicating in some agencies is the best, face-to-face communications can not be replaced by any other form of communicating. The problems associated with face- to-face communications can be overcome on an emergency with adequate training.

Training involves the setting of simple guidelines before an emergency. For example, having all supervisors' report of the incident commander unless advised to do otherwise and instructing subordinate managers who are within a line of sight or a reasonable distance to report in person. It is important that supervisors recognize that when orders are unclear, the objective uncertain, actions to be taken not identified clearly or where safety is a real concern, that face-to-face communication must be utilized.

Generally a subordinate manager should be briefed on a face-to-face basis if authority and responsibility is being released. The plan should be explained, organization outlined, diagrams provided, controls specified, and reference make to what can be seen from the command post.

Performance-oriented communications is what emergency managers utilize. The most effective methods to release authority and responsibility direct activities or explain coordination is face-to-face communications. A network that uses other methods exclusively does not communicate all vital and individually meaningful information.

18 FIRE SITUATIONS THAT SHOUT

"WATCH OUT" - "USE CAUTION"

1. THE FIRE IS NOT SCOUTED AND SIZED UP.
2. YOU'RE IN COUNTRY NOT SEEN IN DAYLIGHT
3. YOUR SAFETY ZONES AND ESCAPE ROUTES AREN'T IDENTIFIED.
4. YOU'RE UNFAMILIAR WITH WEATHER AND LOCAL FACTORS INFLUENCING FIRE BEHAVIOR
5. YOU'RE UNINFORMED ON STRATEGY, TACTICS, AND HAZARDS.
6. INSTRUCTIONS AND ASSIGNMENTS ARE NOT CLEAR.
7. YOU HAVE NO COMMUNICATIONS LINK WITH CREWMEMBERS AND SUPERVISORS.
8. YOU'RE CONSTRUCTING A LINE WITHOUT A SAFE ANCHOR POINT.
9. YOU'RE BUILDING A FIRELINE DOWNHILL WITH FIRE BELOW.
10. YOU'RE ATTEMPTING A FRONTAL ASSAULT ON THE FIRE
11 THERE IS UNBURNED FUEL BETWEEN YOU AND THE FIRE
12. YOU CANNOT SEE THE MAIN FIRE AND YOU'RE NOT IN CONTACT WITH ANYONE WHO CAN.
13. YOU'RE ON A HILLSIDE WHERE ROLLING MATERIAL CAN IGNITE FUEL BELOW.
14. THE WEATHER IS GETTING HOTTER AND DRIER.
15. WIND INCREASES AND/OR CHANGES DIRECTION.
16. YOU'RE GETTING FREQUENT SPOT FIRES ACROSS THE FIRE LINE
17. TERRAIN AND FUELS MAKE ESCAPE TO SAFETY ZONES DIFFICULT ½
18. YOU FEEL LIKE TAKING A NAP NEAR THE FIRELINE.

TEN (10) STANDARD WILDLAND FIREFIGHTING ORDERS

1. KEEP INFORMED ON FIRE WEATHER CONDITIONS AND FORECASTS.

2. KNOW WHAT THE FIRE IS DOING AT ALL TIMES.
      (Personally observe or use lookouts)

3. BASE ALL ACTIONS ON CURRENT AND EXPECTED FIRE BEHAVIOR.

4. HAVE ESCAPE ROUTES FOR EVERYONE AND MAKE THEM KNOWN.

5. POST A LOOKOUT WHEN THERE IS POSSIBLE DANGER.

6. BE ALERT, KEEP CALM, THINK CLEARLY AND ACT DECISIVELY

7. MAINTAIN PROMPT COMMUNICATIONS WITH PERSONNEL, SUPERVISORS AND ADJOINING FORCES.

8. GIVE CLEAR INSTRUCTIONS AND BE SURE THEY ARE HEARD AND UNDERSTOOD.

9. MAINTAIN CONTROL OF PERSONNEL AT ALL TIMES.

10. FIGHT FIRE AGGRESSIVELY BUT PROVIDE FOR SAFETY AT ALL TIMES

FOUR COMMON DENOMINATORS

**FOUR COMMON DENOMINATORS OF FIRE BEHAVIOR
TRAGEDY AND NEAR-MISS WILDLAND FIRES**

1. MOST OF THE INCIDENTS OCCURRED ON RELATIVELY SMALL FIRES OR ISOLATED SECTIONS OF LARGER FIRES

2. MOST OF THE FIRES WERE INNOCENT IN APPEARANCE - IN SOME CASES IN THE MOP-UP STAGE - PRIOR TO THE "FLARE-UPS" OR "BLOW-UPS."

3. FLARE-UPS OCCURRED IN DECEPTIVELY LIGHT FUELS.

4. FIRES RAN UPHILL IN "CHIMNEYS, GULLIES OR ON STEEP SLOPES."

DOWNHILL FIREFIGHTING RULES

FIREFIGHTERS SHOULD NEVER TRY TO BUILD LINE DOWNHILL IN STEEP TERRAIN AND FAST BURNING FUELS, UNLESS THERE IS NO SUITABLE ALTERNATIVE FOR CONTROLLING THE FIRE, AND THEN ONLY WHEN THE FOLLOWING SAFETY REQUIREMENTS ARE ADHERED TO:

A.   THE DECISION IS MADE BY A COMPETENT FIREFIGHTER AND THEN ONLY AFTER THOROUGH SCOUTING.

B.   THE TOE OF THE FIRE IS ANCHORED.

C.   THE FIRELINE DOES NOT LIE IN OR ADJACENT TO A CHIMNEY OR CHUTE THAT COULD BURN OUT WHILE CREW IS IN THE VICINITY

D.   COMMUNICATIONS ARE ESTABLISHED BETWEEN THE CREW WORKING DOWNHILL AND THE CREW WORKING TOWARD THEM, WHICH MAY BE AT THE TOE OF THE FIRE. WHEN NEITHER CREW CAN ADEQUATELY OBSERVE THE FIRE, COMMUNICATIONS WILL BE ESTABLISHED BETWEEN THE CREWS AND A LOOKOUT POSTED WHERE THE FIRE BEHAVIOR CAN BE SEEN.

E.   THE CREW WILL BE ABLE TO RAPIDLY REACH A ZONE OF SAFETY FROM ANY POINT ALONG THE LINE IF THE FIRE UNEXPECTEDLY CROSSES BELOW THEM.

F.   DIRECT ATTACK WILL BE USED WHENEVER POSSIBLE.

G.   IF DIRECT ATTACK IS NOT POSSIBLE; THE FIRELINE SHOULD BE COMPLETED BETWEEN ANCHOR POINTS BEFORE BEING FIRED OUT. FIRING OPERATIONS SHOULD PROCEED WITH ASSURED ACCESS TO THE BURNED-OUT PART OF THE FIRELINE OR OTHER SAFETY ZONES.

H.   FULL COMPLIANCE WITH THE TEN STANDARD ORDERS.

LCES

LCES - A Key to Safety in the Wildland Fire Environment. LCES stands for L - Lookout(s) C - Communication(s) E - Escape routes S - Safety zone(s).

LCES is a system for operational safety in the wildland. In the wildland fire environment are four basic safety hazards that confront the firefighter. They are lighting, fire-weakened timber, rolling rocks, and entrapment by running fires. Together the elements of LCES form a safety system used by wildland firefighters to protect themselves. The safety system is put in place before fighting the fire. Select a lookout or lookouts set up a communications system choose escape routes and select a safety zone or zones.

In operation LCES functions sequentially - it's a self-triggering mechanism Lookouts assess and reassess the fire environment and communicate to each firefighter threats to safety firefighters use escape routes and move to safety zones. Actually all firefighters should be alert to changes in the fire environment and have the authority to initiate communication.

Key Guidelines built into LCES are as follows

❑ Before safety is threatened each firefighter must be informed how LCES will be used.
❑ The LCES system must be continuously reevaluated as fire conditions change.

How to make LCES work.

❑ Train lookouts to observe the wildland fire environment and to recognize and anticipate fire behavior changes.
❑ Position lookout or lookouts where both the hazard and the firefighters can be seen. (Each situation - the terrain cover and fire size determines the number of lookouts that are needed) Remember all firefighters have the responsibility to warn others of the threat to safety.
❑ Set up communications by radio voice or both by which the lookout(s) warn firefighters promptly and clearly of approaching danger. It is paramount that every firefighter receives the correct warning in a timely manner.
❑ Establish the escape routes (at least two)
❑ Reestablish escape routes as their effectiveness decreases.

❑ Establish safety zone locations where the threatened firefighter may find adequate refuge from the danger. A true safety zone should not require deployment of a fire shelter.(This does not imply that one should not be deployed in a safety zone)

The LCES system approach to fireline safety was developed by Paul Gleason USFS Fort Collins, CO after studying fatalities and near misses for over 20 years. LCES simply refocuses on the essential elements of the standard FIRE ORDERS. All wildland firefighters should know LCES W Lookout Communication Escape Routes and Safety zone interconnection.

WILDLAND FIRE ENTRAPMENT FATALITIES

DATE	FIRE	CIVILIAN DEATHS	FIREFIGHTERS BURNED	FATALITY	AGENCY	REMARKS
1825	Main, USA – New Brunswick, Canada	160	Unknown	Unknown	Multiple	Drought
10/1871	Pestigo, Wis.	1300	Unknown	Unknown	Multiple	Largest fatality recorded
1910	Montana-Idaho	7	Unknown	72	USFS Multiple	Fire Storm – Gale Force Winds
1910	Cabinet NF			4	USFS	Same as above
1910	Pend Oreille NF			2	USFS	Same as above
10/12/18	Minnesota	559	Unknown	Unknown	Multiple	Few details obtainable
10/04/22	Ontario Canada	44	Unknown	Unknown	Multiple	
1926	Toiyebe, Kings Cyn	5			USFS	Down Hill/Wind Change
1929	Dellar Mt CA	1			USFS	Uphill Outrun
1933	Kamus Burn	2			USFS	Uphill Outrun
10/3/33	Griffith Park	28	128	25	LA City	W.P.A. Crew
1936	Chatsworth, NJ			5	Other	Wind Shift
1937	Welcome Lake			1	USFS	Crown Fire
1937	Blackwater			15	USFS	Spot Fire
1938	Pepper Run, PA			8	Multiple	Wind Shift
10/11/38	Minnesota	22			Multiple	
1939	Rock Creek, Humbolt			5	USFS	Grass/Unexpected Run
1939	Bixby Mt. Carmel CA			1	CDF	Unknown
1940	Sweetwater, San Diego			1	CDF	Disobedience to orders
1940	Silver Plume, Lincoln			1	USFS	Sudden Wind Change
1941	Kawailoa, Hawaii			2	Multiple	Flashy Fuel – Steep Slope
1943	Williams Hill, Los Padres			1	USFS	Run in Grass
1943	Hauser Creek		Unknown	11	USFS	Cleveland NF No details
1943	Rogerson, ID			2	Unknown	Running brush fire
1944	Hot Springs, Payette			1	USFS	Unknown
1945	Williams Canyon, CA			1	CDF	Unknown
1947	Big Tujunga, LA County		4	4	USFS	Spot Fire
1947	Bryant Canyon, Angeles			2	USFS	Spot Fire Below
1948	Barrett Dam, CA			1	USFS	Wind Change - Night
1949	Walton Spur, CA			1	USFS	Trapped Above Fire
1949	Warm Spr., Payette			1	USFS	Unexpected Winds
1949	Hells Canyon Payette ID			1	USFS	High Winds
1949	Mann Gulch, CO		3	13	USFS	Rapid Spread - Grass
1950	Trabucco, Cleveland CA			1	CDF	Uphill run
1950	Pelitor Fire, CA			4	CDF Military	Unknown
1952	Old Camp, Orange CA			1	CDF	Unknown
1952	Morrell Fire, CA			1	CDF	Unknown
1952	Glenville, Arkansas			1	State	High Winds
1953	Bonnie Blue, Virginia			1	Unknown	Steep Slope
7/9/53	Rattlesnake, CA		15	15	USFS	Spot fire, night wind reversal

DATE	FIRE	CIVILIAN FIREFIGHTERS DEATHS BURNED	FATALITY	AGENCY	REMARKS
1954	Gap Creek, Tennessee		3	State	Slope Above Fire
1954	Tunnel #6, Tahoe, CA		3	USFS	Flare-up at Night
1955	Brickyard, NJ		1	Other	Spot in woods
1955	Johnson AZ		1	USFS	Run Light Fuel
1955	Coyote Los Padres CA	26	1	USFS	Chimney/Saddle
1955	Verdugo, CA	1	1	LA City	Chimney/Saddle
1955	Sagebrush. OR		1	USFS	Cumulus Clouds
9/2/55	Hacienda, CA	10	6	LA Co.	Gas Flashover
1956	East Highlands, CA		1	USFS	Light Fuel - Night
9/1/56	Inaja, Cleveland, CA	Unknown	11	Multiple	Night, flashover, poor Communications(USFS card system)
1958	Stewart, Cleveland CA		1	USFS	Flare-up
1958	Albert Ranch Angeles CA		1	USFS	Flare-up
9/1/58	Tracer, CA	2	1	LA Co.	Spot fire below a ravine
12/25/58	Corral Canyon, CA		8	LA Co.	Spot fire below a saddle
1959	San Luis Obispo, CA		1	CDF	Unknown
1959	Pennington, Texas		1	Unknown	High Winds
1959	Gun, Angeles, CA		1	USFS	Flare-up
8/8/59	Decker, Cleveland, CA	11	7	Multiple	Wind change during backfiring
1959	Stabke, San Bernardino, CA		1	USFS	Unstable Atmosphere
1959	Dry, Sequoia, CA		1	USFS	Down slope Wind
1960	Silver City North Carolina		1	Unknown	Grass Fire
1960	Florida		1	Unknown	Light Fuels - Winds
1960	Georgia		8	Unknown	
1960	Cummings Cr., OR		1	USFS	Wind Change
1961	Silver Creek, Nezperce ID		2	USFS	Spot Fire- Steep Slope
1961	Sierra, Angeles NF, CA		1	USFS	Wind Change
8/2/62	Timber Lodge, Sierra NF, CA	9	4	USFS-CDF	
1964	Don Pedro Fire, CA		1	CDF	Unknown
1964	Joshua Falls, Virginia		1	Unknown	Draw
1964	Coyote, Los Padres , CA		1	USFS	Down slope Wind
1965	Fairview Hollow, Kentucky		3	City	Steep Slope
1965	Helker, N. Carolina		1	Unknown	Wind Speed
11/1/66	Loop, Angeles NF, CA	21	12	USFS	Spot Fire - Below cute (checklist - downhill)
11/3/66	Piedra, Cleveland NF, CA	4	4	USFS	
1967	Baliff, CA	1	1	USFS	Very little info available
1967	Ginnis Lake, WI		1	DNR	Spot – Main fire caught
1967	Range, KS		1	Other	Out run grass fire
1967	Sundance, Kootenai, ID		2	USFS	Major Run - Blowup
1967	Slaughter AZ		1	USFS	Out run Fire
1967	Windsor, S. Carolina		1	Unknown	Dry - Windy

DATE	FIRE	CIVILIAN FIREFIGHTERS DEATHS BURNED	FATALITY	AGENCY	REMARKS
1967	Mississippi		1	Unknown	Gusty winds
1968	Williams CO		1	USFS	Outrun Fire
1968	Ivey, FL		1	Unknown	Overrun by fire
1968	North Carolina #2, NC		1	Other	Clothes caught fire
1968	NC		1	Other	Unknown
10/24/68	Canyon Inn, CA	11	8	LA County	Spot Fire below a 1/2 acre bowl flare-up
1968	Cleveland NF, CA	Unknown	1	USFS	No details available
1970	Shasta-Trinity NF, CA		1	USFS	Smoke Jumper
10/26/70	Fork, Angeles NF, CA	5	5		Helicopter crash
1971	Virginia		1	State	Steep Slope
9/19/71	Mack #2, San Berdo NF, CA	1	1	Multiple	Disregard of 10 Standard Orders
10/7/71	Ramero, Las Padres NF, CA	6	4	Multiple	Inadequate time for escape by tractor. Overran by sundowner
1972	Arkansas		1	Unknown	High Winds - Snag
1972	Idaho		2	Unknown	Thunderstorm
1972	Warthog, Tennessee		1	Unknown	Wind Gusts
8/11/73	Bell Valley, San Diego, CA		1	CDF	
1975	Stockton, UT		2	Other	Wind changed direction
1976	Buhler, KS		1	Other	Tried to outrun fire
1976	Battlement Ck, Colorado		3	Unknown	Steep Draw
1977	Cart Creek, UT		3	USFS	Wind change – steep slope
1977	Okanogan, WA		1	USFS	Uphill spread
1977	New Jersey #1		1	Unknown	Shifting winds in grass
1977	Bass River, NJ		4	Unknown	Crown fire
1977	Vandenburg AFB, CA		3	Ventura County	Erratic winds – 80 mph.
1977	Riverside, TX		1	Other	Winds gusting – 45 mph
1977	Washington #1, WA		1	Fire Dist.	Fire Whirl - Grass
1978	Lost Creek, OR		1	BLM	Slash burning
1978	Terrell Co, GA		1	Other	Pine – Grass 17 mph wind
6/5/79	Sanborn, Santa Clara Co., CA		1	CDF	Heat Stroke - Pneumonia
1979	Oklahoma, OK		1	Other	Wind shifted
1979	Ship Island, ID		1	USFS	Rapidly increasing fire
1979	Okefenokee, FL		1	Other	Wind shift
8/15/79	Spanish Ranch San Louis, CA	4	4	CDF	Chimney
1980	Mack Lack, MI		1	USFS	Swept over by fire
1980	Texas #2		1	Unknown	Grass fire
1981	Merritt Island, FL		2	Other	Winds shifted to 45 MPH.
1981	Elizabeth, Angeles, CA		1	USFS	Fire grew on spot fire
1981	Williams Hill, SC		1	Unknown	Tried to run through fire
1982	Canoe Landing, WI		1	Other	Sudden change in wind direction
1983	Mechlenburg, NC		1	Unknown	Firewhirl

DATE	FIRE	CIVILIAN FIREFIGHTERS DEATHS	BURNED	FATALITY	AGENCY	REMARKS
1983	Nevada #1			1	BLM	Wind shift
1984	Rainbow Springs, AR			2	USFS	Sudden flareup
1984	Clear Lake, CA			1	CDF	Light fuels – direct attack
1984	Texas #3, TX			1	Unknown	Sudden wind change
1985	Kahneetah Lodge, OR		1	2	BIA	Tried to outrun uphill
1985	Sandlin Bay, CO			1	Unknown	Dozer caught by fire
1985	Golden Gate Ests., FL			1	DNR	3 heads – over run by fire
1985	Adams Run, SC			1	Unknown	Increase in spread rate-intensity
1987	Lake Co, CA			1	CDF	Rapid fire progression
1988	Blue Hole #6, OR			1	DNR	Contractor entrapped – No shelter deployed
1989	Buckhorn, OH			1	DNR	Wind shift – sudden increase in fire
6/25/90	Dude Fire, Tonto NF AZ		6	6	USFS	Steep Slope, Dry Fuel, Low Humidity
1990	California Fire, CA		3	2	CDF	Shifting winds – extreme temp. – 17 shelters deployed
1990	Wasatch State Park, UT			2	State	Winds increased
1990	Grand Coulee, WA			1	Fire Dist.	Tried to outrun fire
1991	Neon, KY			1	Pvt.	Sudden flare up
10/16/91	Spokane, WA			1	Multiple	
10/18/91	Oakland, CA	26	Unknown	1	Multiple	Power Pole Fell
1993	Angles NF, CA			2	LA County	Spotting above crew
1993	Santa Fe NF, NM			1	USFS	Localized wind event 40-50 mph.
1994	Augusta, WI			1	Unknown	Large wind gust
1994	Hull Mt., Medford, OR			1	USFS	Contractor – Burn over dozer
1994	Cedar Fire, GA			1	USFS	Erratic fire behavior
7/6/94	Storm King – So. Cyn. Fire - Colorado			14	USFS	Spot Fire - Steep Slope – Uphill Run
1994	Tennessee State			1	Pvt.	Fire change direction
1995	Point Fire ID			2	Volunteer	Sudden wind change – Firefighters entrapped in vehicle – heavy fuels
10-26-96	Calabasas, Coral Cyn., CA		6		LA County	Extreme rate of spread
1996	Costrip, MT			2	Pvt.	Fire change direction – tried to outrun grass fire
7/10//01	Twisp, WA		1	4	Mulitple	Left Safety of Vehicle

ATTACKING THE WILDLAND/URBAN INTERFACE FIRE

Many communities today are faced with growing urbanization in areas that interface with primary watershed and forestlands. Even rural agencies are experiencing unparalleled growth as people try and escape urban strife. This growth has brought a host of new firefighting problems to fire departments and districts that previously have had to deal only with occasional small incidents of this nature. As cities and districts continue annexation and grow into these wildland/urban interface areas, they must become more proficient in meeting a different set of situations calling for a departure from conventional or normal firefighting tactics.

Even local state and federal fire control personnel are often not the best source for preparing your personnel to deal with initial attack situations. This is because in many areas of the country they have a different mindset when it comes to initial fire attack. In fact in some areas they are more restrained in their approach to wildland fires because of the distances they must travel, apparatus design, and/or type and size of the incident when they arrive. They are prone to taking a less aggressive attack posture, waiting for additional resources before Initiating tactical operations on active fire lines, and making absolutely certain that their command structure is in place with all safety issues considered.

Let me make it clear, these agencies for the normal type of problem they are faced with are dealing with the situations in an appropriate manner. This is because they are attacking with wildland fires that are often times beyond what most cities and districts consider the initial stages. The reason this occurs is units have half-hour or longer arrival times and minimum forces are sent until such time as the first unit provides a report on conditions. Cities and districts because of risks have stations that are closer, generally send more apparatus, and thus attempt to suppress the fire very quickly.

Generally, the local city and fire district problems center around two primary elements, recognition by officers and tactical considerations.

Recognition by Officers

Even large fire departments with numerous companies experienced in wildland firefighting have been caught short-handed during major emergencies in populated urban interface areas. This generally occurs due to the failure of officers to recognize the significance of the problem. No agency has sufficient resources to meet every demand. Officers must recognize that any fire in or near fringe areas has the potential of quickly becoming a major conflagration. Additional companies, task forces, strike teams, etc., need to be ordered long before the problem escalates beyond the capabilities of the agencies' own resources.

Pre-determined staging areas should be given consideration before an emergency occurs. When a mental commitment of companies by an officer exceeds the resources available from within the department, a call should go out for mutual aid units to respond to these staging areas. This is can be critical during periods when rapid escalation of the situation is occurring.

If pre-determined staging areas are not identified make sure you identify your staging area upon your request for additional resources. The agency may want to consider the use of some of your personnel report to staging to guide companies not familiar with the geographical area.

Tactical Considerations

Wildland urban interface areas during emergencies fall into broad tactical categories of initial attack and major situations.

Initial Attack

When a company is first due, the officer in charge needs to make some quick tactical decisions. The fire isn't going to remain static; therefore the company must be committed to some form of action. Generally the officer should base a decision on what course of action best minimizes the problem. Usually an attack on the flank or head of the fire which poses the greatest potential for impact on the area's structures is best. This may mean that the fire will become larger which means you have "traded errors." As an officer you have made a known tactical mistake by attacking a slower moving area (the flank vs. the head) to protect the buildings in that area, thereby allowing the fire to grow.

One tactical mistake that should not be made is to commit in an initial attack situation to protecting structures while allowing the Incident to escalate, impacting more buildings. This may mean In the early stages when resources are most critical, a building which is already afire may have to be passed by for the time being to prevent a conflagration;

Use of Engines

An engine shou1d remain mobile working with water available in the tank, using smaller (10-50) gpm nozzles {the best nozzle has a selective range of 10 to over 100 gpm) and no hose larger than 1-1/2" to stretch operating time. Units with four-wheel off road capability will most likely be one of your best offensive weapons. These units when possible need to work inside the burn area and should never be parked in the green or unburned area next to an active fire line. Attacks on the head of the fire from the green must be avoided at all costs.

The most effective attack is a mobile with personnel on hose lines walking out front of the vehicle on the ground, not riding on the engine. This method puts more water on the fire with less waste, better results, and less re-ignition. A mobile attack needs to start from an established anchor point on the flanks, not the head of the fire. When possible, use two lines off one engine or use engines in tandem with one following up the first or primary attack unit. When personnel are available, the attack lines should be followed up with hand tools as a part of your mobile attack action.

Conventional drive or structural engines may not have direct access to the fire line or be capable of a mobile attack. Long 2-1/2" or larger hose lays to water sources should be avoided. Tenders or shuttle operations are usually the most effective ways to deliver water.

Engines should be parked in a location so as not to impede the flow of traffic, especially of mobile attack units. Progressive hose lays of 1-1/2" hose should be used from an anchor point. This may be possible to pass through the green safely at the rear of the fire to reach a flank to start an attack. It is not wise to pass through the green to attack an active flank, and never do so to attack the head.

Bulldozers

Bulldozers (dozers) can be an effective initial attack tool especially in fuels such as grass and medium brush. Dozers like engines need to start their attack from an anchor point. Generally this means from the rear or the point of ignition or along a flank, again like engines, never attack the head or through the green. Dozers in an initial attack situation are used much like a hose line or mobile attack on the fire line only in this case with half the blade in the burn and half in the green or unburned area. When practical, if you have a four-wheel engine available following the dozer to suppress hot spots may be advisable. The traditional dozer boss used by some agencies is not necessary, but is a good idea if personnel are available to have one person on the ground to work with the operator.

Dozers are classified as light, medium, and heavy. A .light unit is a D-4 (50 hp). A medium dozer is a D-5 or 6 (100 hp) and a heavy unit is a D-7 (200 hp) or larger. These units with their blades angled toward the burn area on one pass can cut a fire line at an average of 700 yards per hour for light dozers, 850 yards per hours for medium dozers, and 1000 yards per hour for heavy dozers... These amounts may be higher on down grades vs. upgrades and will certainly be more on flat grassland as well as reduced in heavier vegetation. Dozers can work on slops of up to better than 50%, but is usually wise to avoid such applications.

If your agency is not large enough to have a dozer and/or an operator you may want to consider contracting with a local equipment company. Many business of this type are community minded and will provide such services at no or minimal costs for the first hour or two. Your may want to discuss any liability concerns with your agencies legal council prior to entering into any type of agreement and make certain you provide sufficient training.

All personnel in initial attack situations should have proper safety equipment, including wildland protective clothing, lightweight helmets, and appropriate gloves as well as boots, and a fire shelter. Personnel need to be trained in the standard wildland firefighting orders and situations that shout watch out and know how to use their protective equipment and tools correctly. Safety must always be a concern and this is the primary purpose of working from the burn area or with one foot in the burn and one in unburned fuel or green. The safety zone or escape route is just a step away in the burn or black area.

Major Emergencies

Just when a problem is a major wildland emergency may be unclear to an officer not familiar in dealing with wildland/urban interface fires. Generally, a major emergency should be declared when a fire becomes too complex to handle. (For instance, when it is beyond the capabilities of the first alarm assignment to stop the fire from spreading to other buildings.) This type of situation calls for a shift in tactical operations. Instead of concentrating on the wildland fire, major emphasis is placed on protection of structures. The keys are water supply, resources and mobility.

Water supply

Conventional firefighting tactics in urban areas rely heavily on fire hydrants for water. During a wildland interface fire, conservation and effective use of water carried on the units may be of more importance than the water system depending on design. One thing should be made clear, tactically, the system might fail.

The best system is gravity feed from a water source above the area being impacted. Although, if the problem is of major significance areas below the water source located on foothills may be without effective water pressure. This occurs because of heavy demand from homeowners wetting down roofs in lower areas not being impacted by the fire. Tactical planning to avoid these situations should include the water company's knowledge of where to shut down key valves to maintain water pressure when and where needed.

Water systems that rely on pumps to provide pressures are generally even more unreliable during major fires. Pumps often have a greater problem keeping up with demands. Should they be electric, power supplies may be lost; especially if overhead wires that are subject to burning supply the power. The power company should be made aware of the importance of electricity in supplying water during major emergencies. Sometimes they have standing orders to cut service when a problem reaches certain proportions.

Companies working in an area should be prepared to work entirely from water carried on the unit. Whenever possible, even garden hoses from buildings should be used for small fires on exposures.

Resources

The largest impact on any department during a major fire may be the resources available or lack thereof. One thing is certain; all departments can ill afford to waste a single company or person. Tough tactical decisions may have to be made by every officer. When periods of high winds contribute to the fire's potential spread, certain buildings may have to be written off in order to allow resources to save others. Generally buildings sitting in saddles and at the top of draws are poor risks that may require the commitment of too many companies to save. The same is true of homes in box canyons with one access.

Buildings which have already started to bum should be considered lost. This is especially true of residences where wood shingle roofs are involved and the fire is taking off upon arrival of a company. Regardless of the situation, officers should be made aware of when they are on their own without additional backup capabilities. This should help avoid making commitments to operations they themselves cannot accomplish. As the amount of resources increase or the problem de-escalates, officers should be advised to start committing to more fire involved situations especially when additional resources can be provided if needed.

Chief officers in charge of the overall operation should assign resources to cover a given size area depending on projected or anticipated problems. The more units available or the more severe the problem the smaller the areas should become that each unit is expected to cover.

Mobility

The key to any major wildland urban interface fire is a high degree of mobility. Officers must be prepared to move with the fire. They must be trained to recognize when the area has become less active and the major threat has passed so they can move on to meet more critical needs.

Generally, commitment to hydrants should be avoided if resources are in short demand. Fire hose can be run right off of hydrants with smaller gpm nozzles. They can be left in an area when the fire passes. Companies should avoid picking up hose once it is laid.

Wetting down of roofs generally does little good in a high wind situation, unless large quantities *of* water are available for a sustained period *of* time. But be cautious, large flows may create even more critical needs in other areas. Foam and gel type solutions may be practical when available but it is usually wise to wait to apply such products to buildings just prior to impact for maximum effectiveness. Companies should be trained to enter a building and remove combustibles from around windows on the fireside. Garden hoses should be used where practical to put out small fires on roofs. Axes should be used to chop away burned areas on wood shingle roofs. This officer has personally seen a shingle company, armed with only ladders, axes and wet sheets, save more homes with wood shingle roofs than two other companies committed to hydrants on the same street.

Engine companies should be set up with 150 to 200' length of 1-3/4" or 1-1/2" hose, coiled and charged on top of the hose bed so it can be pulled any direction for quick utilization. Hose lays of 2-1/2" or larger hose need to be avoided, as do lays of more than one line or lays over a distance of 200'. Engines should be backed in when protecting structures so the can be used as a safety zone if necessary when the fire approaches. Always have an escape route for the unit and never block the escape route of another piece of equipment.

Summary

As officers, we will not always be able to meet every demand and certainly we will be subject to public criticism no matter what our course of action we take. The only way to minimize this impact is to provide the necessary training for our personnel and coordinate with other support agencies such as law enforcement, water and power companies, etc. In addition, we should develop public information programs that will make all concerned aware of the problems and how we are prepared to deal with them.

GANSER HOSE PACK

(2) 100-FOOT ROLLS OF 1-1/2 INCH SINGLE JACKET HOSE IN A HOSE PACK USED FOR WILDLAND AND HIGH RISE FIREFIGHTING.

THIS HOSE PACK WILL INCLUDE TWO BUNDLES OF 1-1/2 INCH HOSE. THE FIRST BUNDLE WILL HAVE ARM SLINGS AND THE SECOND BUNDLE WILL BE SECURED TO THE FIRST. EACH WILL HAVE 400 FEET OF 4-4/2INCH HOSE FOR A GRAND TOTAL OF 200 FEET. THE BUNDLE MAY INCLUDE A NOZZLE, ADAPTER, WYE, LATERAL, OR OTHER FITTINGS AS NECESSARY.

FIRST PACK: START WITH THE MALE END. ROLL HOSE IN A THREE-FOOT CIRCLE; WHEN COMPLETELY ROLLED, PUSH THE FEMALE INTO THE CENTER. ALLOW FOR ARM SLINGS AND SECURE *USING STRONG, LIGHT ROPE (SUCH AS PARACHUTE CORD) WITH A SLIPKNOT. USE A PAIR OF PLIERS TO MAKE THE KNOT EXTRA TIGHT.*

SECOND PACK: REPEAT THE ABOVE INSTRUCTIONS, OMITTING THE ARM SLINGS AND TYING THE SECOND PACK TO THE FIRST PACK AS ILLUSTRATED.

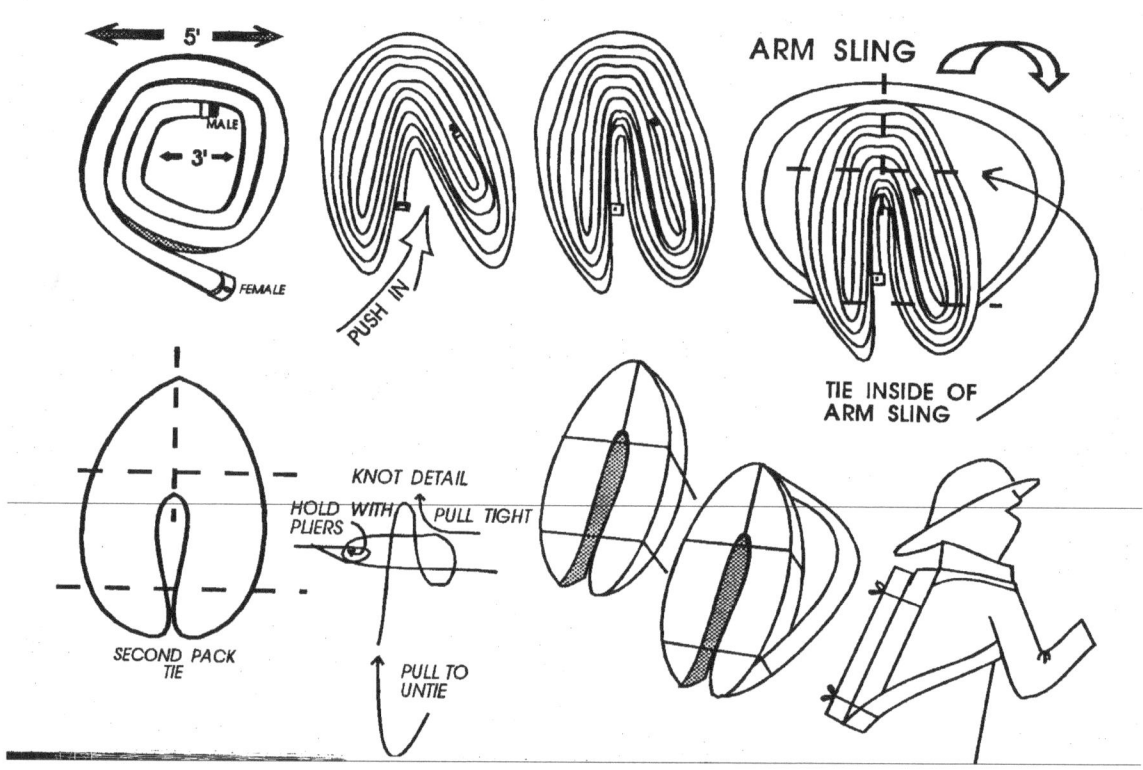

GANSER HOSE PACK R IS CHARGED ON TUE GROUND WITHOUT UNROLLING - JUST UNTIE DROP TO TUE GROUND - UNDO TUCK (PUSH IN) TO FORM A CIRCLE AND CHARGE.

GROUP PROBLEM #1

Your company has been assigned to protect a house in a wildland/urban interface area of numerous homes. As you enter the area the fire blocked your only means of egress. The house with a wood shake roof is located on a slight uphill slope. The wind is blowing at 3540 mph, relative humidity is less than 20%, and the temperature is about 90 degrees. It is estimated that the fire will impact the house in about 30 minutes.

PROBLEM 1

TYPICAL WILDLAND/URBAN STRUCTURE PROTECTION ASSIGNMENT

300 FEET TO

Identify the steps you can take to increase the chances of this house avoiding destruction. One option you do not have is to leave the structure.

GROUP PROBLEM #2

THE FIRE IS APPROACHING A TRACT OF HOMES – WIND 25-30 MPH, RELATIVE HUMIDITY LESS THAN 20%, AND THE FUEL IS LIGHT GRASSES, MIXED WITH MEDIUM BRUSH AS WELL AS MIXED CONFER TREES. MOST OF THE HOME HAVE WOOD SHAKE ROOFS. YOUR ENGINE AND ONE OTHER HAVE BEEN ASSIGNED BY YOUR STRIKE TEAM LEADER TO PROTECT THESE STRUCTURES. THE FIRE WILL IMPACT YOUR POSITION IN LESS THAN 20 MINUTES. IDENTIFY WHAT HOMES YOU WOULD PROTECT AND EXPLAIN THE ACTIONS YOU WOULD TAKE.

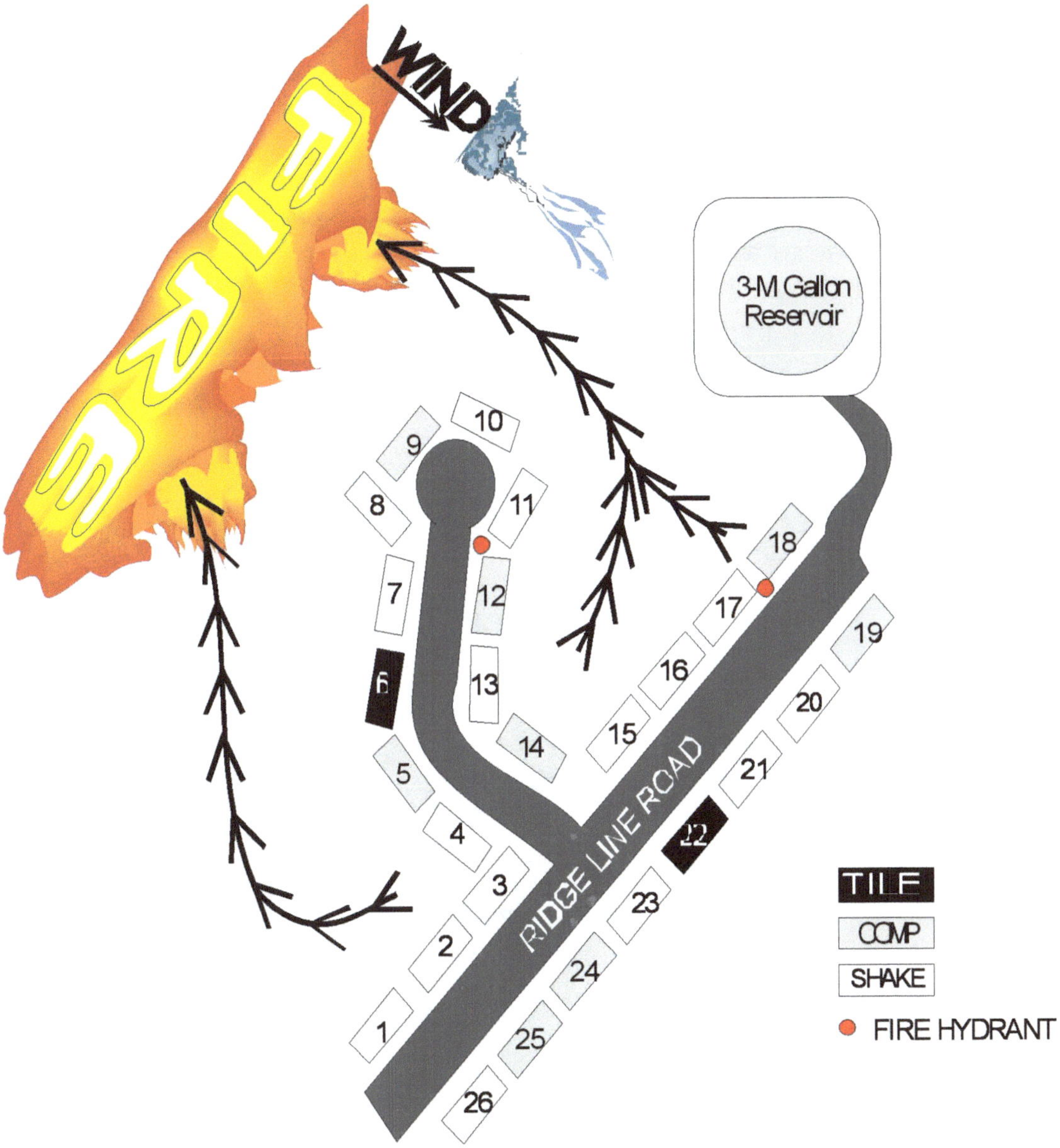

WIND

FIRE

3-M Gallon
Reservoir

9
10
8
11
7
18
12
17
6
13
16
5
14
15
19
4
20
3
21
2
22
23
1
24
25
26

RIDGE LINE ROAD

TILE
COMP
SHAKE
● FIRE HYDRANT

RESERVOIR

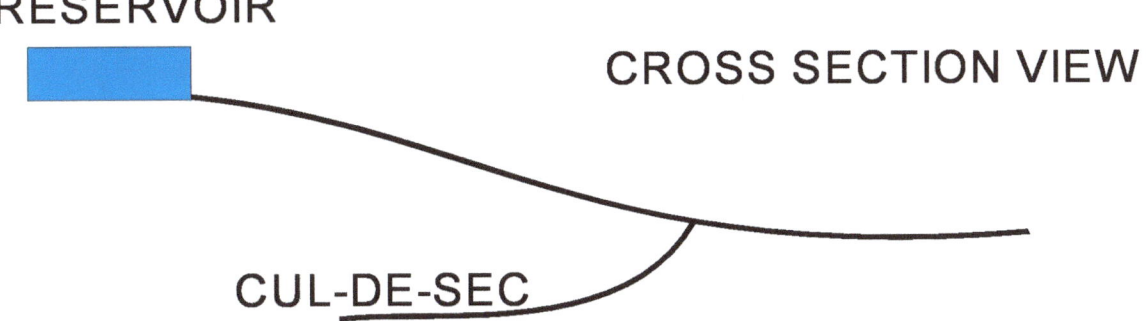

CROSS SECTION VIEW

CUL-DE-SEC

GROUP PROBLEM #3

Your company has been assigned to protect a house in a wildland/urban interface area of numerous homes. As you enter the area the fire blocked your only means of egress. The house with a wood shake roof is located on a slight uphill slope. The wind is blowing at 3540 mph, relative humidity is less than 20%, and the temperature is about 90 degrees. It is estimated that the fire will impact the house in about 30 minutes.

House with swimming pool and greenbelt

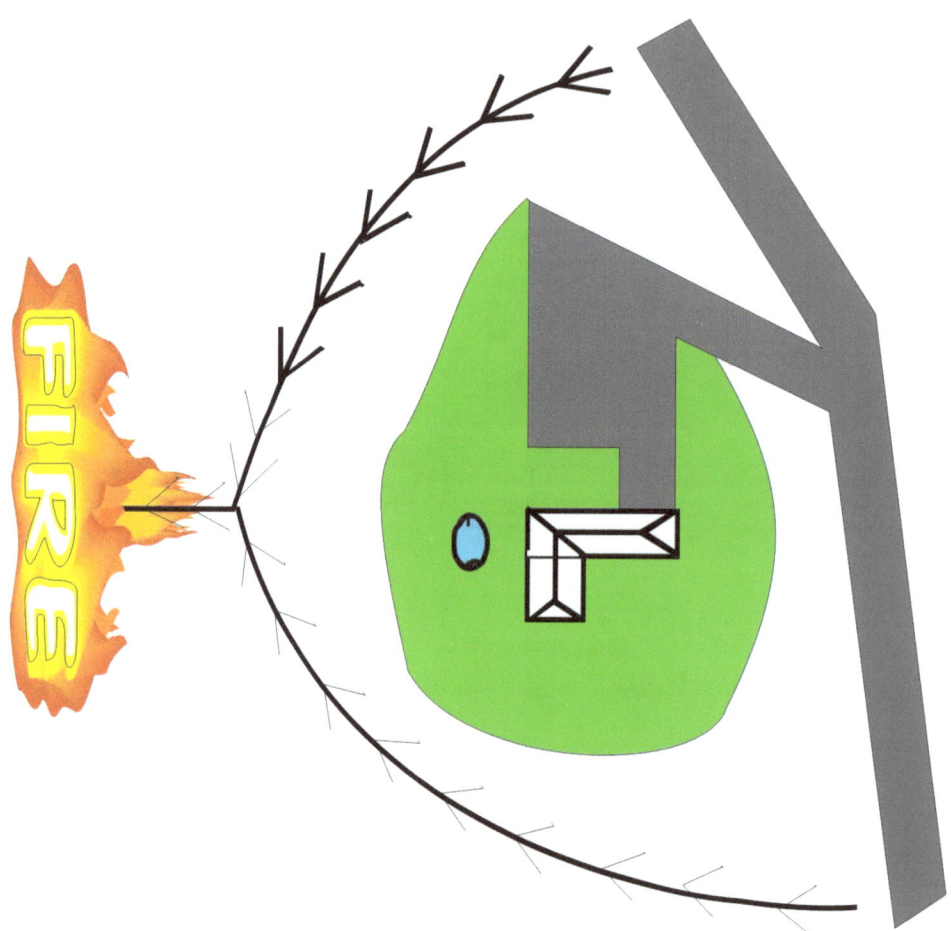

 Identify the steps you can take to improve safety for you, your apparatus, and still protect the building from loss. One option you do not have is to leave the structure.